Handbook of Food Contamination and Safety

Handbook of Food Contamination and Safety

Emmett Norton

MURPHY & MOORE
www.murphy-moorepublishing.com

Murphy & Moore Publishing,
1 Rockefeller Plaza,
New York City, NY 10020, USA

Visit us on the World Wide Web at:
www.murphy-moorepublishing.com

ISBN: 978-1-63987-280-0 (Hardback)

Cataloging-in-Publication Data

Handbook of food contamination and safety / Emmett Norton.
p. cm.
Includes bibliographical references and index.
ISBN 978-1-63987-280-0
1. Food contamination. 2. Food--Safety measures. 3. Food adulteration and inspection.
I. Norton, Emmett.
TX531 .H36 2022
363.192 6--dc23

Table of Contents

Preface

It is with great pleasure that I present this book. It has been carefully written after numerous discussions with my peers and other practitioners of the field. I would like to take this opportunity to thank my family and friends who have been extremely supporting at every step in my life.

The presence of harmful microbes and chemicals in food which can degrade its quality, is referred as food contamination. It can cause foodborne illnesses such as food poisoning, listeriosis, salmonellosis, etc. The common types of contamination include biological, physical, chemical and cross-contamination. The scientific discipline of preparation, handling and storage of food in order to reduce foodborne illnesses is known as food safety. The main aim of food safety is to prevent food from getting contaminated using several methods to maintain its quality. Methods for ensuring food safety include food labeling, certification systems of food, guidelines for management of food export and import, maintaining food hygiene and regulation of food additives and pesticide residues. This book is a compilation of chapters that discuss the most vital concepts in the field of food contamination and safety. The topics covered herein deal with the core aspects of food contamination and safety. This book will serve as a reference to a broad spectrum of readers.

The chapters below are organized to facilitate a comprehensive understanding of the subject:

Chapter - Food Contamination

Food contamination refers to the presence of some unwanted chemicals and harmful microorganisms in the food which causes food-borne illnesses and food allergies. Some of the names of the food contaminants are nitrosamines, polycyclic aromatic hydrocarbons (PAH), heterocyclic amines, histamine, acrylamide, etc. This chapter discusses the aspect of food contamination in detail.

Chapter - Types of Food Contamination

There are mainly four types of food contaminants which are biological, chemical, physical and cross contamination. These contaminants are mainly for safety and quality concerns. This chapter has been carefully written to provide an easy understanding of the various types of food contaminants.

Chapter - Food Preservation

The food can be preserved by using many preservation techniques such as, pickling, canning, pasteurization, food drying, biopreservation, fermentation, chemical preservation, low-temperature preservation, thermal and non-thermal processing of food, etc. This chapter discusses these varied food preservation techniques in detail.

Chapter - Food Safety

The practices which aim at preserving the quality of food and prevent it from contamination and food borne illness during preparation, handling and storage are referred to as food safety. Safe food handling, frozen foods, food hazard analysis, food packaging, etc. are some of its concepts. The topics elaborated in this chapter will help in gaining a better perspective of food safety.

Chapter - Foodborne Illnesses

The consumption of contaminated food can cause many foodborne illnesses. Some of them are diarrhea, gastroenteritis, traveler's diarrhea, vibrio vulnificus, food poisoning, etc. This chapter closely examines these different foodborne illnesses to provide an extensive understanding of the subject.

Emmett Norton

1

Food Contamination

Food contamination refers to the presence of some unwanted chemicals and harmful microorganisms in the food which causes food-borne illnesses and food allergies. Some of the name of the food contaminants are nitrosamines, polycyclic aromatic hydrocarbons (PAH), heterocyclic amines, histamine, acrylamide, etc. This chapter discusses the aspect of food contamination in detail.

SHELF LIFE OF FOOD

The shelf life of food is the period during which the food retains an acceptable quality from a safety and organoleptic point of view, and depends on four main factors, namely formulation, processing, packaging and storage conditions. Foods are perishable and there are many factors that can deteriorate the quality and safety of food products during storage and distribution. These can be categorized into chemical and physical factors. To minimize the degradation of food during processing or storage, kinetic models which describe degradation rates and the dependence of intrinsic factors on extrinsic factors, such as temperature and moisture content, must be determined. The essential purpose of kinetic models is to describe sufficiently a set of experimentally obtained data. These kinetic models can then be used for prediction, process control, optimization and simulation of various food processing operations. It allows what-if scenarios and insights into the food systems to be explored and simplifies the process design for maximizing shelf life. Shelf life models are mathematical equations which describe the relationship between the food, the package and the environment. These models are based on different degradation factors (i.e., critical factors) and are essential in predicting the shelf life of food, in designing the packages, and providing useful insights about the foods under abnormal circumstances along the supply chain. Most of the efforts in terms of mathematical modelling have focused on microbiological safety and spoilage.

Manufacturers are also challenged with determining and maximizing the shelf life for products that are exposed to varying conditions in the supply chain. Shelf life touches on all the issues mentioned, and shelf-life determination is an essential requirement in providing safe, quality food products to consumers.

There are many definitions of shelf life provided by governments and organizations. The Institute of Food Science and Technology defines shelf life as "the period of time during which the food product will remain safe; be certain to retain its desired sensory, chemical, physical, microbiological, and functional characteristics; where appropriate, comply with any label declaration of nutrition data, when stored under the recommended conditions." Both food safety and quality are important aspects of acceptable shelf life. Although pathogens are usually monitored during shelf-life studies, a suitable food safety program is the best way to ensure the product's safety.

Factors affecting Shelf Life

Both intrinsic and extrinsic factors influence the shelf life of food products.

Intrinsic factors include the following:

- Initial quality: For perishable food, the initial microbial load will influence the shelf life. Using ingredients that have already started to deteriorate (e.g. old oil) or over processing can result in loss of texture or nutrients (e.g. vitamin C).
- Inherent nature of the product: Fresh or perishable foods have an inherently shorter shelf life than shelf-stable foods. The low water activity of a product such as rice makes it an inherently shelf-stable food, for example.
- Product formulation: The addition of preservatives or antioxidants can extend the shelf life of the product. Formulation changes such as replacing the type of acid, removing nitrates from a processed meat, and reducing the amount of added salt can also change the shelf life of the product.

The following are extrinsic factors:

- Processing methods: Thermal processing will reduce (e.g. pasteurization) or eliminate (e.g. sterilization) microbes and extend the shelf life of the product. Other gentle processing techniques such as high pressure processing can also be used to reduce initial microbial levels.
- Packaging: For shelf-stable products, the barrier of the package can affect the shelf life. For example, moisture absorption for a cracker will affect the crispness of the product and a moisture barrier is required. If the product has a large fat component (e.g. potato chips), fat oxidation affects the shelf life and an oxygen barrier is required. Light protection may also be required. Without light protection, milk is susceptible to vitamin degradation and off-taste due to light-induced oxidation.
- Transportation and storage conditions: Exposure of the product to variable temperatures and relative humidity in the supply chain (including the retail environment) can affect the shelf life of foods. For refrigerated products,

higher-than-optimal temperature storage can accelerate microbial growth. Oxidation reactions are also accelerated by higher temperature exposure, thus shortening the shelf life of products.

- Consumer handling: After purchase, transfer of food from the store to home can result in higher temperature exposure. Consumer refrigerators can also be at higher-than-optimal storage temperatures. Once the package is opened, the shelf-life date assigned by the food manufacturer is no longer applicable.

Understanding the End of Shelf Life

What constitutes the end of shelf life? The end point can be indicated from relevant food legislation, guidelines provided by government or professional organizations, or the use of acceptable industry practices. Often acceptability limits are chosen based on self-determined end points. For the most part, the food industry relies on sensory perception as an indicator of product failure. Product acceptability may be determined when there is a significant difference in the aging sample compared to a fresh sample by using discrimination testing (e.g. paired comparison, triangle, duo-trio, etc.). Descriptive analysis with expert panelists describes the change in sensory attributes (e.g. odor, taste, appearance, and texture) and can indicate consumer rejection. Although acceptance testing or use of consumer panels for acceptability can be more accurate, it is seldom used since a large number of panelists are required, resulting in a more time-consuming and expensive process.

FOOD CONTAMINATION

To start with, bacteria are present in the animals raised for food. Meat and poultry can become contaminated during slaughter through cross-contamination from intestinal fecal matter. Similarly, fresh fruits and vegetables can be contaminated if they are washed using water contaminated with animal manure or human sewage. During food processing, contamination is also possible from infected food handlers. Lastly, poor hygiene in the home is also a factor.

Bacterial Food Contamination

Many bacteria can contaminate food. The most common include the following:

- Campylobacter jejuni: Mishandling of raw poultry and consumption of undercooked poultry are the main causes of C. jejuni contamination.
- Clostridium botulinum: Bacteria producing a toxin in food responsible for botulism, the deadly paralytic nerve illness.
- Escherichia coli: A leading cause of food contamination. Based on a 1999 estimate, 73,000 cases of infection and 61 deaths occur in the United States each year. The E. coli 0157: H7 strain is found in ground beef, raw milk, chicken, vegetables, and fruit.
- Salmonella typhimurium: Salmonella contamination can occur in meats, poultry, eggs or milk products.
- Shigella: The most common food that these bacteria can contaminate include: salads (potato, chicken, seafood, vegetable), raw vegetables, milk and other dairy products, and meat products especially poultry.
- Staphylococcus aureus: Can be found in custard or cream-filled baked goods, ham, poultry, eggs, potato salad, cream sauces, sandwich fillings.
- Vibrio cholera: These bacteria cause the well-known disease cholera that has caused many outbreaks all over the world. It can be transmitted by water or food.
- Vibrio vulnificus: Free-living ocean bacteria that can cause food borne illnesses from contaminated seafood. Especially dangerous in the warm weather months when eating shellfish that are undercooked or raw.

Spoiled milk is also mostly caused by bacteria such as Lactococcus cremoris or Enterobacter aero-genes, that cause the milk to form long white strands.

Water contamination is usually due to the presence of three bacteria, E. coli, Clostridium perfringens, and enterococci, the bacteria normally found in the feces of people and many animals.

Parasitic Food Contamination

Parasites are organisms that lives in or on a host, and obtain nourishment without benefiting or killing the host. They enter the body through the mouth when contaminated food or drink is swallowed. There are many different types and range in size from single-celled, microscopic organisms (protozoa) to larger, multi-cellular worms (helminths) that can be seen without a microscope. Parasites that contaminate food include:

- Entamoeba histolytica: Parasite that causes amoebic dysentery, characterized by severe diarrhea. It is transmitted by contaminated water, and is often called "traveler's dysentery" because of its prevalence in developing nations.
- Giardia duodenalis: Microscopic parasite that can live in the intestines of animals and people. It is found in every region of the world and is one of the most common causes of waterborne and foodborne) illness.
- Cryptosporidium parvum: Microscopic parasite, a significant cause of water contamination worldwide. It is found in the intestines of many herd animals including cows, sheep, goats, deer, and elk.
- Cyclospora cayetanensis: Single-celled, microscopic parasite. Little is known about this organism, although cases of infection are being reported from various countries with increasing frequency.
- Toxoplasma gondii: Single-celled, microscopic parasite found throughout the world. Found in foods such as raw or undercooked meats, especially pork, lamb, or wild game, and in drinking untreated water.
- Trichinella spiralis: Intestinal roundworm whose larvae may migrate from the digestive tract and form cysts in various muscles of the body. In the United States, infections are most prevalent where pork or wild game is consumed raw or undercooked.
- Taenia saginatajsolium: Taenia saginata (beef tapeworm) and Taenia solium (pork tapeworm) are parasitic worms (helminths).

Food Contaminants

A food contaminant is a substance, object, or organism that makes food unfit for human consumption. The contamination of the food may have occurred by accident or on purpose. Some dictionaries and authorities say that the term only includes substances that got into food by mistake. A food contaminant may also be something that somebody added to a foodstuff to make it bulkier or weigh more. Somebody may have added water to milk, for example, to increase its volume.

A food contaminant might be harmless food that is not in the right place. Carrots, for example, may be in a bag of frozen broccoli.

Examples of substances or organisms that can make food unfit for human consumption are insects, pests, dust, or biocides. Animal feces and pathogenic microorganisms may also make food unfit to eat. Pathogenic means it can cause disease.

Food contaminants are any harmful substances unintentionally added to food, which may be chemicals from natural sources, environmental pollution, or formed during food processing".

Food Contaminant Types

Biological

A biological food contaminant is either an organism or a substance that an organism produces. This includes biological matter that microorganisms, insects, rodents, and humans produce.

The two main causes of biological contamination are viruses and bacteria. They can cause some of the most common foodborne illnesses.

A foodborne illness is an illness that exists because of something we ate.

"Thoroughly washing your hands and sanitising (US: sanitizing) the food handling equipment are two of the best ways to prevent against bacterial contamination."

Physical

A physical contaminant is a foreign object in food, i.e., it should not be there. Physical contamination may occur at any stage of the production process.

Pieces of plastic, glass, and steel wool, for example, may contaminate food.

The main risk for people who eat foods with physical contaminants in them is injury. However, some injuries or undesirable events are life-threatening. For example, if a piece of plastic gets stuck in your throat, you might not be able to breathe.

If a physical contaminant is carrying a biological contaminant, there is a risk of disease or infection.

Chemical

Chemical contaminants are either artificial or natural substances that got into food. In some cases, the chemical is toxic, i.e., poisonous. Toxins can make us ill and even kill us.

The impact of a food contaminant on consumer well-being and health is often not apparent for many years. It can take decades before we become aware that anything is wrong.

Unlike biological contaminants, thermal processing does not affect many chemical contaminants. Thermal processing is a commercial technique to sterilize food by heating it.

Below are some examples of chemical food contaminants:

Agrochemicals

Agrochemicals are chemicals that farmers use, such as herbicides, insecticides, and rodenticides. Plant growth regulators and veterinary drugs are also examples of chemicals that farmers use.

Environmental Contaminants

Environmental contaminants include chemicals that are present in the environment where we grow food. They may also exist in the environment where we process, package, store, transport, and consume food.

The food may become contaminated when it comes into contact with its environment. Arsenic and mercury, for example, are environmental contaminants that exist in water.

Radionuclides, such as cesium-137 and polycyclic aromatic hydrocarbons are present in the atmosphere.

Perchlorates, cadmium, and nitrates are contaminants that exist in the soil. Bisphenol A and sanitizing agents exist in packaging materials and processing/cooking equipment respectively.

FOOD SPOILAGE

Food is considered contaminated when unwanted microorganisms are present. Most of the time the contamination is natural, but sometimes it is artificial. Natural contamination occurs when microorganisms attach themselves to foods while the foods are in their growing stages. For instance, fruits are often contaminated with yeasts because yeasts ferment the carbohydrates in fruits. Artificial contamination occurs when food is handled or processed, such as when fecal bacteria enter food through improper handling procedures.

Food spoilage is a disagreeable change or departure from the food's normal state. Such a change can be detected with the senses of smell, taste, touch, or vision. Changes occurring in food depend upon the composition of food and the microorganisms present in it and result from chemical reactions relating to the metabolic activities of microorganisms as they grow in the food.

Types of Spoilage

Various physical, chemical, and biological factors play contributing roles in spoilage. For instance, microorganisms that break down fats grow in sweet butter (unsalted butter) and cause a type of spoilage calledrancidity. Certain types of fungi and bacteria fall into this category. Species of the Gram-negative bacterial rod Pseudomonas are major causes of rancidity. The microorganisms break down the fats in butter to produce glycerol and acids, both of which are responsible for the smell and taste of rancid butter.

Another example occurs in meat, which is primarily protein. Bacteria able to digest protein (proteolytic bacteria) break down the protein in meat and release odoriferous products such as putrescine and cadaverine. Chemical products such as these result from the incomplete utilization of the amino acids in the protein.

Food spoilage can also result in a sour taste. If milk is kept too long, for example, it will sour. In this case, bacteria that have survived pasteurization grow in the milk and produce acid from the carbohydrate lactose in it. The spoilage will occur more rapidly if the milk is held at room temperature than if refrigerated. The sour taste is due to the presence of lactic acid, acetic acid, butyric acid, and other food acids.

The general sources of food spoilage microorganisms are the air, soil, sewage, and animal wastes. Microorganisms clinging to foods grown in the ground are potential spoilers of the food. Meats and fish products are contaminated by bacteria from the animal's internal organs, skin, and feet. Meat is rapidly contaminated when it is ground for hamburger or sausage because the bacteria normally present on the outside of the meat move into the chopped meat where there are many air pockets and a rich supply of moisture. Fish tissues are contaminated more readily than meat because they are of a looser consistency and are easily penetrated.

Canned foods are sterilized before being placed on the grocery shelf, but if the sterilization has been unsuccessful, contamination or food spoilage may occur. Swollen cans usually contain gas produced by members of the genus Clostridium. Sour spoilage without gas is commonly due to members of the genus Bacillus. This type of spoilage is called flat-sour spoilage. Lactobacilli are responsible for acid spoilage when they break down the carbohydrates in foods and produce detectable amounts of acid.

Among the important criteria determining the type of spoilage are the nature of the food preserved, the length of time before it is consumed, and the handling methods needed to process the foods. Various criteria determine which preservation methods are used.

References

- Shelf-life-of-foods, food-science: sciencedirect.com, Retrieved 14 February, 2020
- Determining-product-shelf-life: foodqualityandsafety.com, Retrieved 18 May, 2020
- Food-contaminant, financial-glossary: marketbusinessnews.com, Retrieved 22 June, 2020
- Food-contamination: diet.com, Retrieved 16 April, 2020
- Food-spoilage, study-guides-biology-microbiology-food-microbiology: cliffsnotes.com, Retrieved 26 March, 2020

2

Types of Food Contamination

There are mainly four types of food contaminants which are biological, chemical, physical and cross contamination. These contaminants are mainly for safety and quality concerns. This chapter has been carefully written to provide an easy understanding of the various types of food contaminants.

BIOLOGICAL CONTAMINATION

Biological contamination is when bacteria or toxins contaminate food and is a common cause of food poisoning and food spoilage.

Food poisoning can happen when harmful bacteria, also called pathogens, spread to food, and are consumed. Bacteria are small microorganisms that split and multiply very quickly. In conditions ideal for bacterial growth, one single-cell bacteria can split so many times that in just seven hours, it has multiplied into two million.

Some bacteria such as salmonella, staphylococcus and listeria are extremely toxic by themselves. And, sometimes it's not the bacteria that are toxic to humans, but the process of the bacteria multiplying and producing waste. However, not all bacteria are harmful to humans; many are quite beneficial, such as those found in yoghurt.

As a Food Handler, it's your job to control the spread of harmful bacteria by maintaining food safety. Bacteria can be found everywhere and are impossible to see with the naked eye. Some of the most common places for bacteria to grow are:

- The human body,
- Dust,
- Raw meat,
- Pets and pests,
- The air,

- Kitchen cloths,
- Food Handler's clothing.

How Bacteria Survive?

- Food: Bacteria need a constant source of food to survive, especially protein. High protein foods such as meat are particularly vulnerable to biological contamination from bacteria, which means they're considered high-risk foods.
- Water: Water is essential to bacterial growth and without it, most bacteria will die. Which is why drying foods as a way of preservation are so effective and have been performed for thousands of years.
- Oxygen: Most bacteria require air to survive, these are called aerobic bacteria. Although some bacteria - called anaerobic bacteria - can survive without oxygen. Which is why it's still possible to get food poisoning from canned food items.
- PH Levels: PH refers to food acidity and is measured on a scale of 1 (acidic) to 14 (alkaline). Most fruits generally have a PH level of between 1 - 5.9, so are considered acidic. While many alkaline foods such as vegetables have a PH level at the other end of the scale. Bacteria thrive in neutral foods that are neither acidic or alkaline and generally have a PH level of between 6 - 8.9. Foods such as meat and seafood are prime examples of neutral foods.
- Time and temperature: Bacteria need both time and the right temperature to multiply to dangerous levels. A temperature of between 5 °C and 60 °C - also referred to as the 'danger zone' - allows for maximum bacterial growth, so it's important not to keep food at this temperature for too long.

High-risk foods are those that have ideal conditions for bacterial growth. This means they're usually:

- Neutral in acidity,
- High in starch or protein,
- Moist.

Foods such as seafood, cooked rice or pasta, and dairy are all considered high-risk because they provide the perfect environment for bacteria to grow. This is why it's essential to practice proper food handling when dealing with these foods.

Low-risk foods are those that don't have particularly good bacterial growth conditions. These foods are:

- High in acidity,
- High in salt or sugar,

- Dried,
- Canned or vacuum packed.

Some examples of low-risk foods would be pickles, uncooked rice or pasta and jams. Although these foods are not common sources of biological contamination, the appropriate care must still be taken when handling them.

PHYSICAL CONTAMINATION

Physical contamination happens when actual objects contaminate foods. Sometimes when a food is physically contaminated, it can also be biologically contaminated. This is because the physical contamination might harbour dangerous bacteria, for example, a fingernail.

Common sources of physical contamination are:

- Hair: Always wear hair neatly tied back and use a hair net if possible.
- Glass or metal: This can occur when kitchen items are not maintained. Cracked or broken crockery and utensils should be thrown away, as well as any food that might have come into contact with it.
- Pests: Pests - such as mice, rats and cockroaches - leave droppings that can contaminate food. Also, pests themselves - such as flies and insects - can also make their way into food.
- Jewellery: Always keep jewellery to a minimum when preparing and handling food.
- Dirt: Because dirt is so small, it's easy not to notice that it's contaminating your food. It usually gets into the food from unwashed food and vegetables.
- Fingernails: Always keep nails short and clean to prevent contamination. Also, avoid wearing fake nails as these can fall off and may contaminate food.

Most raw foods and ingredients originate in a natural environment such as a field, an orchard or a farm. As the food is harvested, foreign objects such as stones or glass can end up commingled and transported into the processing plant.

When a crop is ready to be harvested, a tractor-like machine actually grabs the trunk of each tree and vigorously shakes it so the almonds fall to the ground. The almonds are then left on the ground for several days to dry out; another machine then sweeps them into rows so a harvester can pick them up with a series of belts. They are then cleaned through the machine and dumped into a bucket in the back of the truck. The truckload of almonds is then taken to the huller and sheller plant where additional debris, shells and sticks are removed, and the nuts are cleaned and sorted via several pieces of heavy machinery so only the kernels remain. Sometimes rocks, pieces of metal or glass particles are found and removed through the sorting and cleaning process.

As the food moves into the processing and packaging facility, there is potential for more foreign physical contaminants. The food production industry is run on machinery that can break down and wear out. As a result, sometimes small pieces of that machinery can end up in a product or package. Metal contaminants can be accidentally introduced in the form of loose pieces of equipment such as nuts, bolts and washers, or metal can break off from other parts such as mesh screens and filters.

How to Detect Physical Contaminants

The food industry takes many precautions to ensure that any food that reaches consumers is free of physical contaminants like metal, glass and stone, which as we just learned can enter a product or package anytime during farming or processing. There are X-ray inspection systems that use X-rays and sensors and food metal detectors that use metal coils and high frequencies to find contaminants before they reach consumers. In fact, this year there have been major changes in food metal detection technology that will help significantly in reducing the number of metal physical contaminants in food manufacturing. Food metal detectors that are equipped with new multiscan technology enable operators to pick a set of up to five frequencies from 50 kHz to 1,000 kHz. The technology then scans through each frequency at a very rapid rate, effectively acting like five metal detectors in one.

Investigation and Identification of Physical Contaminants

Most physical contaminants of foods, such as pieces of hard plastic or wood, can cause consumers immediate injury; this includes all types of foods, including beverages, bottled water, and nutritional and functional products. Any physical material in food that does not belong in the product may be classified as a physical contaminant.

Physical contaminants are also referred to as physical hazards or foreign matter. Shrew teeth in a crop product or a piece of wire in a meat product are examples. Glass pieces, metal fragments, bone chips, and pits may all cause serious harm when ingested. Common examples of bodily harm include lacerations of the lips, the inside of the mouth, teeth, gums, tongue, throat, esophagus, stomach, and intestine, and even choking. Children and seniors are at greater risk and have the highest incidence of such harm.

Government agencies, food producers, manufacturers, distributors, and retailers must all protect the health of the consumer as one of their most important objectives and responsibilities. If there is any evidence or reason to believe there are physical contaminants in a food product that may cause illness or injury to consumers, government agencies, reject the food product from sale market or request food recalls.

Huge economic losses to businesses may occur when a physical hazard is discovered. There were 456 food recalls in 2017 and 764 food recalls in 2016, which were announced by FDA and FSIS. Of those, foreign matter caused 42 recalls in 2017 and 44 recalls in 2016.According to FSIS, 16.51 million pounds of food were withdrawn from the market following 56 food recalls because of extraneous materials from 2015 to 2017.

Table: Causes of Physical Contamination.

Sources	Example of containments
Field	Rocks/ stones/ sand, asphalt, metals/ bullets, concrete particles, bones, wood fragments, and thorns.
Processing	Glass, ceramic/shards, metal fragments, staples, blades, clips, needles, keys, screws, magnet fragments, washer bolts, screening, plastic, grease/lubricants, rubber, insulation/seal materials, nail polish, jewellery, coins, pieces of gloves, finger cots, bandages, cigarette butts, gum, bones, pits, fruit stones, nut & animal shells, medications/ tablets/ capsules, wood, pens, and pencils.
Storage and distribution	Metal, plastic, and wood fragments.

Physical contaminants in food could come from either external sources, such as metal fragments, or internal sources, such as bone particles and pits. They can be introduced into food products accidentally during harvesting or at any point during processing due to poor procedural practices anywhere in the food chain, including manufacturing, storage, transportation, or retail. The so-called Dirty Dozen, the 12 most common foreign material contaminants in food, are glass, wood, stones, metal, jewelry, insects/ filth, insulation, bone, plastic, personal effects, bullets/BB shot, and needles. Table provides a summary of common sources.

Food companies work hard to keep their products free of contaminants. Investigation and control of physical contaminants in food should be conducted throughout the whole processing chain or in food testing laboratories. With accurate and timely information, a thorough investigation can be carried out in testing laboratories in a cost-effective and efficient manner. It is essential to have a professional investigation team with appropriate resources and equipment to help food company quality assurance staff troubleshoot consumer complaints and answer questions as to what the contaminants and their sources are. It is often necessary to apply integrated, multidimensional approaches for complicated investigations.

Comprehensive investigation can demand microscopy-based examinations plus chemical techniques, on-site examinations, Fourier transform-infrared (FTIR) analysis, and other techniques. The investigation processes and screening procedures depend on the particular physical contaminants and their sources. The procedures can be combined or modified methods found from the following: AOAC International, American Spice Trade Association, FDA's Macroanalytical Procedures Manual, FDA's Laboratory Information Bulletins, USDA's foreign matter identification documents, the United States Pharmacopeia, and other compendious sources.

In general, special investigation procedures include:

- Inspect and target suspected sources of physical contaminants,
- Identify the foreign matter,
- Determine or evaluate the root cause or sources.

These procedures consist of eight steps:

- Target contaminants: In many cases, it is necessary to select a suitable investigational procedure, including sample inspection, preparation, and identification of target contaminants. The sources of contamination are diverse, whether individual or related, or even unknown. Often, little is known about the contaminants, how many there might be, what size, what their regulatory status might be, or whether they might be in a food mixture.

  Pre-examination steps allow judgments to be made as to whether the targets can be separated by size, shape, mass or magnetic properties, the type of samples and contaminants (e.g., organic or inorganic), amount, circumstances of contamination, and levels. It is normally required to conduct resampling, inspection, macroscopic examination, extraction, filtration, floating, sieving, burning, dyeing, and further examination using X-ray, metal detectors/magnets, and other screening or targeting technologies. An example might be to use magnets to gather and identify ferrous metal particles from liquid samples.

- Identify contaminants: Once the suspected foreign matter screening procedures are carried out using stereo or dissecting microscopic examinations as a starting point, one can obtain detailed evidence, identifying the morphology and deciding what methods of analysis should be applied. For example, if a complainant believes that pieces of glass were observed by the naked eye in a food, it may be that the material consists of rocks, salt, sugar, plastics, minerals, struvite, or tartrate crystals. If needed, a compound microscope, bright-field/dark-field microscope, polarized microscope, or scanning electron microscope could be applied to reveal more details. Further tests are determined based on the results of microscopic examinations. Selection of such analytical strategies may require a combination of techniques or the development of multidisciplinary approaches, depending on contaminant conditions and the goals of the investigation.
- Conduct physical property tests: A polarized microscope is used to display birefringent properties from some materials, such as synthetic polymers. Spectroscopic techniques can reveal specific functional groups of a chemical material. For example, different plastics may be identified by FTIR analysis. Physical properties can be classified by many characteristics and features, such as size, shape, thickness, magnetic characteristics, solubility, buoyancy, elasticity, flexibility, flammability, temperature resistance, etc.

- Perform a chemical examination: Chemical analysis, including an elemental analysis, can reveal characteristic features that enable an understanding of chemical properties. Histochemical staining techniques are useful to test the chemical and physical properties of the contaminants. For instance, chemical reactions can be used for determination of lignin characteristics and enzymatic reactions.
- Confirm: In many cases, no single method is 100 percent guaranteed to complete an investigation and identification. Further testing may be required to validate the findings or disprove them. For example, analysis by atomic absorption spectroscopy or other elemental analysis techniques reveals different types of metals. Examination of the combustion qualities of a sample can be used to find and quantitate foreign ferrous metal particles. Protein quantitative tests can confirm the presence of animal matter.
- Compare: Building a reference library is necessary for a forensics laboratory. Identification relies on the availability of good reference library texts and official methods containing authentic reference materials as well as spectral databases to obtain definitive confirmation of the contaminants.
- Evaluate the root cause: Whether a contaminant went through a specific processing step and was introduced into the product prior to or after packaging, potential sources could be uncovered through a prudent review of all test results and existing factors. If a piece of glass was confirmed, the following question will be raised: Is it more likely to be from a lightbulb, bottle, window glass, or drinking glass? This evaluation can be done to obtain more detailed information and evidence from the contaminant's size, shape, mass, and characteristic features, especially compared with authentic reference materials. Any reference samples provided by clients are helpful to address the root cause.
- Prepare a comprehensive report: After the investigation and identification are complete, a comprehensive report should be prepared, containing a summary of the project goals, sample information, test methods, imaging evidence, findings, evaluations, and suggestions.

CHEMICAL CONTAMINATION

Chemical contamination occurs when food comes into contact with chemicals and can lead to chemical food poisoning.

Some common sources of chemical contamination can include:

- Kitchen cleaning agents: Proper storing of kitchen cleaning chemicals is essential. Never keep food stored in the same place as your cleaning chemicals, and always use cleaning products designed especially for kitchen use.

- Unwashed fruits and vegetables: The pesticides and fungicides often used on fruits and vegetables to help them grow free from diseases are harmful if consumed. Which is why it's vital to properly wash all fruits and vegetables before eating them.
- Food containers made from non-safe plastics: Single-use items - such as plastic containers - are not designed to be reused again and again. Always store food in containers that are specially designed to safely be reused.
- Pest control products: Items like fly spray and rat poison are extremely hazardous if consumed. Always store these products away from food items.
- Chemicals used in equipment maintenance: Some kitchen machines and equipment with moving parts - such as slicers and mixers - can need regular oiling. Always use food-safe oil to help make sure this doesn't contaminate the food you use them to prepare.

Analysis of Chemical Contaminants in Food

Chemical contaminants may occur in our food from various sources. They typically pose a health concern, resulting in strict regulations of their levels by national governments and internationally Therefore, analysis of relevant chemical contaminants is an essential part of food safety testing programs to ensure consumer safety and compliance with regulatory limits. Modern analytical techniques can determine known chemical contaminants in complex food matrices at very low concentration levels. Moreover, they can also help discover and identify new or unexpected chemical contaminants.

Sources of Chemical Contaminants in Food

Chemical contaminants can be present in foods mainly as a result of the use of agrochemicals, such as residues of pesticides and veterinary drugs, contamination from environmental sources (water, air or soil pollution), cross-contamination or formation during food processing, migration from food packaging materials, presence or contamination by natural toxins or use of unapproved food additives and adulterants.

Pesticide Residues

The use of pesticides, such as insecticides, fungicides or herbicides, has become an integral part of modern agriculture to increase crop yields and quality by controlling various pests, diseases and weeds. Registration of new pesticides is a strictly regulated process that evaluates their toxicity and environmental fate, and sets maximum residue limits (tolerances) in raw and processed commodities. There are over 1,400 known pesticides. Some of them should no longer be used but may still be present in the environment. Older pesticides are being reevaluated based on currently available scientific data.

Approved uses of pesticides following Good Agricultural Practices should result in pesticide residues below maximum residue limits established in a given country. However, global sourcing of raw commodities and global distribution of food products complicate the situation because pesticide registrations, uses and limits can be and are different in different countries. Consequently, an approved use in one country may result in an illegal pesticide residue in a food imported into another country, such as the recent case of the fungicide carbendazim in orange juice imported into the United States from Brazil. Furthermore, pesticides can be misused or present in food due to contamination during application (spray drift), storage or transportation or from environmental sources, such as contaminated water or soil.

Veterinary Drug Residues

Similar to pesticides, veterinary drugs are agrochemicals that undergo a thorough registration process, resulting in setting of their maximum residue limits/tolerances in animal-derived foods. The major classes of veterinary drugs include antibiotics, anthelmintics, coccidiostats, nonsteroidal anti-inflammatory drugs, sedatives, corticosteroids, beta-agonists and anabolic hormones. These drugs, which are administered to live animals, can remain as residues in animal tissues. Liver and kidney are highly susceptible to residues given their biological function.

Certain antibiotics, such as penicillin, can cause severe allergic reactions in sensitive individuals, which is an important reason for enforcing their residue limits in foods of animal origin. Another important justification for limiting antibiotic usage in food-producing animals is to reduce the risk of pathogenic microorganisms becoming resistant to antibiotics. Most veterinary drugs are not of acute toxicological concern, but some substances, such as nitrofurans, chloramphenicol, clenbuterol and diethylstilbestrol, have been banned in most countries due to their carcinogenicity. Concern about endocrine-disrupting effects has become another reason for regulation of certain veterinary drugs, such as beta-agonists and hormones.

Environmental Contaminants

Environmental contaminants can be man-made or naturally occurring substances present in air, water or soil. They can enter the food chain and even bioaccumulate. Some can pose an acute health risk if present at higher concentrations, but the major concern related to the presence of environmental contaminants in foods is their potential endocrine disruption, developmental, carcinogenic and other chronic effects.

Examples of environmental contaminants that enter the food chain include heavy metals, polychlorinated biphenyls (PCBs), “dioxins” (polychlorinated dibenzodioxins and dibenzofurans), persistent chlorinated pesticides (e.g., DDT, aldrin, dieldrin, heptachlor, mirex, chlordane), brominated flame retardants (mainly polybrominated diphenyl ethers), polyfluorinated compounds, polycyclic aromatic hydrocarbons (PAHs), perchlorate, pharmaceutical and personal care products or haloacetic acids and other water disinfection byproducts.

The manufacture and use of PCBs and other persistent organic pollutants (POPs) have been banned for years, but they remain in the environment due to their high stability. PAHs can be found in the environment as a result of industrial pollution or can originate from oil spills; thus, they were of concern in seafood after the oil spill accident in the Gulf of Mexico in 2010.

Food Processing Contaminants

Certain toxic or undesirable compounds can be formed in foods during their processing, such as during heating, baking, roasting, grilling, canning, hydrolysis or fermentation. Precursors of these contaminants can occur naturally in the food matrix, such as in the case of acrylamide being formed during the Maillard reaction between the amino acid asparagine and a reducing sugar (especially in potato- and cereal-based, heat-treated products). Alternatively, certain processing contaminants, such as nitrosamines, can be formed by interaction of natural food components with food additives. Carcinogenic and genotoxic chlorpropanols, such as 3-monochloropropane-1,2 diol (3-MCPD), are formed during the acid hydrolysis of wheat, soya and other vegetable protein products.

Examples of other processing contaminants include PAHs (in grilled and smoked products), ethyl carbamate (in yeast-fermented alcoholic beverages and other products) or furan (in a variety of heat-treated foods, especially coffee and canned/jarred food).

Food processing may also be a source of cross-contamination, such as contamination of nonallergenic foods with known food allergens.

Migrants from Packaging Materials

Direct contact of foods with packaging materials can result in chemical contamination caused by migration of certain substances into foods. Examples of migrants of health concern may include bisphenol A or phthalates from plastic materials, 4-methylbenzophenone and 2-isopropylthioxanthone from inks, mineral oil from recycled fibers or semicarbazide from a foaming agent in the plastic gaskets that are used to seal metal lids to glass packaging.

Toxins

Toxins are naturally occurring substances that are produced by various organisms, with mycotoxins and marine biotoxins typically representing the major concerns in foods. Other examples of toxins in foods may include bacterial toxins (e.g., staphylococcal toxins) or certain plant toxins, such as pyrrolizidine alkaloids that can be found in honey, milk or eggs. While the bacterial/fungal contamination can be eliminated with heat treatment, the toxins can remain in the food product as contaminants.

Mycotoxins are toxic secondary metabolites produced by fungi (molds) that can colonize various crops. They are of concern mainly in cereals, nuts, infant formula, milk, dried fruit, baby food, coffee, fruit juice and wine. There are many mycotoxins, but only a few are currently regulated, with the European Union having a more comprehensive list than most other countries, which includes aflatoxins, ochratoxin A, patulin, deoxynivalenol, zearalenone, fumonisins and T-2/HT-2 toxins. Different mycotoxins are prevalent in different climates and in various growing and storage conditions.

Marine biotoxins, such as saxitoxin, domoic acid, okadaic acid or ciguatoxin, are highly toxic compounds produced by phytoplankton. During so-called harmful algal bloom events, they can accumulate in fish or shellfish, such as clams, mussels, scallops or oysters, to levels that can pose serious health risks or even be lethal to humans.

Unapproved Food Additives and Adulterants

Food adulteration can happen accidentally when unapproved additives are introduced to the food, or the wrong additive is introduced through formulation error. This results in mislabeled food. Perhaps a larger health issue is when foods are adulterated intentionally for economic reasons to sell a low-value food or material for more or to mask food spoilage. Some adulteration may just mislead or cheat consumers, such as adding high fructose corn syrup to honey, but some may be harmful to them. The most notorious example from recent years is the addition of melamine to whey and other protein concentrates to increase their apparent protein content analyzed as total nitrogen. Other examples include the use of toxic Sudan dyes in adulterated chili powders or adulteration of virgin olive oil with hazelnut oil, which can cause unexpected allergic reactions in sensitive individuals.

Analysis of known Chemical Contaminants in Food

Most known chemical contaminants in foods are small organic molecules. Except for high-level adulterants, they are typically present in foods at low concentrations (parts per trillion to parts per million); thus, their analyses in complex food matrices are often quite challenging. The basic analytical approach involves an extraction using a suitable solvent, cleanup to remove interfering matrix components, a chromatographic separation and a selective detection.

It is not an exaggeration to say that the implementation of mass spectrometry (MS) as a detection technique has truly revolutionized the analysis of chemical contaminants in foods. As opposed to element-selective or nonselective detectors, MS can detect a wide range of compounds independent of their elemental composition and provide simultaneous quantitation and structural identification of detected analytes. It also adds another degree of separation/selectivity on top of chromatographic separations. These unique features have made MS the number one choice for detection and identification/confirmation of trace-level organic chemical contaminants in modern testing laboratories.

First, the combination of MS with gas chromatography (GC-MS) has become popular for the analysis of volatile and semivolatile compounds, including many pesticide residues, PAHs, PCBs and other less-polar POPs. More polar, thermolabile and less volatile analytes were difficult to analyze until the more recent introduction of atmospheric ionization techniques, such as electrospray, for liquid chromatography-mass spectrometry (LC-MS). LC-MS has opened the door to the direct analysis of many more polar contaminants, including modern, new-generation pesticides, and the majority of veterinary drugs and toxins, such as mycotoxins. Many of the emerging and recently identified contaminants, including acrylamide, melamine or Sudan dyes, are analyzed preferably by LC-MS.

Thus, modern food contaminant testing laboratories utilize both GC-MS and LC-MS to cover the wide polarity range of possible organic chemical contaminants. Tandem MS (MS/MS) is typically employed to provide an increased selectivity (especially in LC-MS) that helps further distinguish target compounds from potential matrix interference. Figure shows an example of chromatograms obtained in a multiresidue analysis of more than 300 pesticides analyzed by GC-MS/MS and LC-MS/MS, demonstrating the speed and selectivity of state-of-the-art instruments that enable the simultaneous and highly sensitive analysis of many compounds.

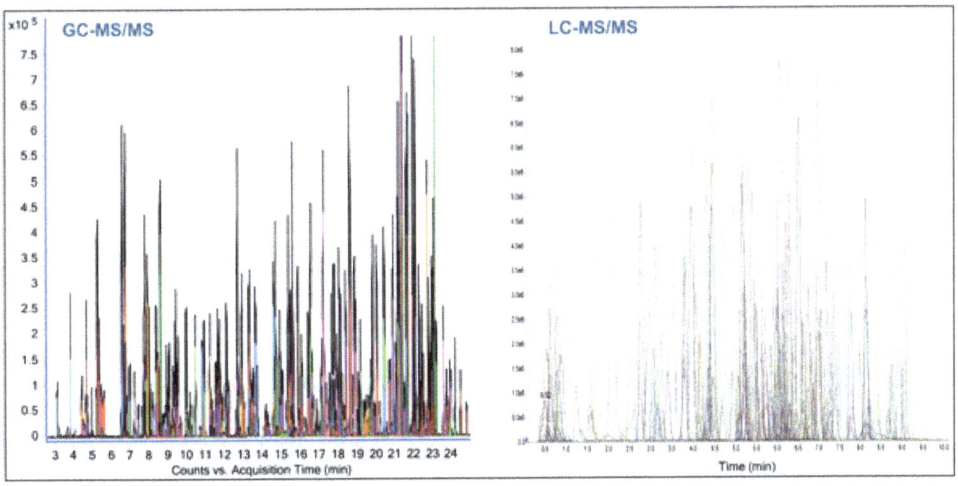

Chromatograms obtained in multiresidue analyses of more than 300 pesticides Analyzed by CG-MS/MS and LC-MS/MS.

Identification of Unknown Chemical Contaminants in Food

Detection and identification of unknown contaminants is not an easy task, especially if they are present at low concentration levels. It requires expertise and a good analytical strategy that is based on all gathered information about the sample and potential sources of contamination. Any clues, such as changes in smell, taste or texture, as well as a description of potential poisoning symptoms may be important in this respect. Concurrent analysis of control ("good") samples with suspect samples is often essential to find differences and eliminate potential false positives.

If a certain compound or a group of compounds is suspected, then a targeted sample preparation and instrumental methods can be employed. For a truly unknown analysis, different extraction and separation approaches should be used to isolate compounds with a wide range of physicochemical properties (polarity, solubility, volatility, etc.). Nontargeted analysis should be performed, such as MS with full-spectra acquisition. Statistical analysis of the acquired chromatographic and MS data of contaminated and noncontaminated samples may help identify differences and reduce the number of components that have to be examined. The acquired MS spectra of suspected contaminants can be compared with MS spectral libraries and compound databases. In LC-MS, high-resolution/accurate-mass measurements, using time-of-flight (TOF) or orbitrap MS instruments, should be used for added selectivity. In addition, tandem MS should be employed to help elucidate the structure of unknown contaminants. In the end, strong knowledge and expertise in both analytical and food chemistry are typically required to succeed in this task.

The current and future trends in the analysis of chemical contaminants are and will be strongly affected by developments in analytical instrumentation. The speed, sensitivity and selectivity of state-of-the-art MS instruments enable analysis of many compounds in one analytical run. Consequently, streamlined sample preparation approaches, such as QuEChERS (Quick, Easy, Cheap, Effective, Rugged and Safe), can be used that require minimum extract cleanup without any preconcentration steps—thus, they can be miniaturized and automated.

The advancements in high-resolution/accurate MS instruments and development of related software tools show great potential to bring this technology from the research environment into testing laboratories, where it could be employed for nontargeted testing of known and unknown chemical contaminants.

The Importance of Moisture in Samples Prior to Chemical Analyses

Moisture content is an important consideration during sampling procedures, in part because it affects the extent of sample heterogeneity. It may be necessary to determine the moisture content through sample drying to express analytical results on a uniform scale.

Precautions must be considered when drying foods at elevated temperatures, since chemical reactions such as hydrolysis can occur and these reactions can be accelerated. Moisture determinations can be erroneous if hydrolysis has occurred, since the water of hydrolysis has not been released from the sample. A general rule of thumb for sample drying is that it should be as rapid and at as low a temperature as possible. Vacuum methods that can used to dry a sample include vacuum ovens and lyophilization, or freeze-drying. Another method is microwave drying. Unlike external heating devices that operate through the sample vessel, microwaves rapidly heat the sample, keeping temperature gradients to a minimum.

For certain chemical testing applications, such as the presence of metals in food as well as the extraction of crude fat and nutrients from food, microwave digestion is the preparation method of choice. The ability of microwave sample prep to dissolve almost any matrix, leaving target species behind, provides preparative capabilities unavailable through other methods.

The inherent variability in the composition of raw materials, basic ingredients and processed foods requires the use of proper sampling and sample pretreatment techniques, in addition to statistical methods for obtaining representative and replicate samples. Using the proper sample preparation methodology can reduce analytical error and costly detection mistakes that could jeopardize the safety of the food produced as well as lead to an even more costly food safety-related recall.

ADULTERATION

Adulteration of food commonly defined as "the addition or subtraction of any substance to or from food, so that the natural composition and quality of food substance is affected". Adulteration is either intentional by either removing substances to food or altering the existing natural properties of food knowingly. Unintentional adulteration is usually attributed to ignorance's, carelessness or lack of facilities for maintaining food quality. Incidental contamination during the period of growth, harvesting, storage, processing, transport and distribution of foods are also considered.

"Adulterant" means any material which is or could be employed for making the food unsafe or sub-standard or mis-branded or containing extraneous matter.

Food is declared adulterated if:

- A substance is added which depreciates or injuriously affects it.
- Cheaper or inferior substances are substituted wholly or in part.
- Any valuable or necessary constituent has been wholly or in part abstracted.
- It is an imitation.
- It is colored or otherwise treated, to improve its appearance or if it contains any added substance injurious to health.
- For whatever reasons its quality is below the Standard.

Adulterated food is dangerous because it may be toxic and can affect health and it could deprive nutrients essential for proper growth and development.

Common Adulterated Foods

Some of the common adulterated foods are milk and milk products, atta, edible oils, cereals, condiments (whole and ground), pulses, coffee, tea, confectionary, baking powder, non - alcoholic beverages, vinegar, besan and curry powder.

Types of Adulterants

Type	Substances Added
Intentional Adulterants	Sand, marble chips, stones, mud, other filth, talc, chalk powder, water, mineral oil and harmful colour.
Incidental adulterants	Pesticide residues, droppings of rodents, larvae in foods.
Metallic contaminants	Arsenic from pesticides, lead from water, effluent from chemical industries, tin from cans.

Poisonous or Deleterious Substances

Generally, if a food contains a poisonous or deleterious substance that may render it injurious to health, it is adulterated. For example, apple cider contaminated with E.coli O157:H7 and Brie cheese contaminated with *Listeria monocytogenes* are adulterated.

If a food contains a poisonous substance in excess of a tolerance, regulatory limit, or action level, mixing it with "clean" food to reduce the level of contamination is not allowed. The deliberate mixing of adulterated food with good food renders the finished product adulterated

Filth and Foreign Matter

Filth and extraneous material include any objectionable substances in foods, such as foreign matter (for example, glass, metal, plastic, wood, stones, sand, cigarette butts), undesirable parts of the raw plant material (such as stems, pits in pitted olives, pieces of shell in canned oysters), and filth (namely, mold, rot, insect and rodent parts, excreta, decomposition.

Economic Adulteration

A food is adulterated if it omits a valuable constituent or substitutes another substance, in whole or in part, for a valuable constituent (for instance, olive oil diluted with tea tree oil); conceals damage or inferiority in any manner (such as fresh fruit with food coloring on its surface to conceal defects); or any substance has been added to it or packed with it to increase its bulk or weight, reduce its quality or strength, or make it appear bigger or of greater value than it is (for example, scallops to which water has been added to make them heavier).

Microbiological Contamination and Adulteration

The fact that a food is contaminated with pathogens (harmful microorganisms such as bacteria, viruses, or protozoa) may, or may not, render it adulterated. Generally, for ready -to-eat foods, the presence of pathogens will render the food adulterated. For example, the presence of Salmonella on fresh fruits or vegetables or in ready-to-eat meat or poultry products (such as luncheon meats) will render those products adulterated.

Ready-to-eat meat and poultry products contaminated with pathogens, such as Salmonella or *Listeria monocytogenes*, are adulterated. For raw meat or poultry products, the presence of pathogens will not always render a product adulterated (because raw meat and poultry products are intended to be cooked and proper cooking should kill pathogens).

ENVIRONMENTAL CONTAMINATION OF FOOD

Environmental contaminants include substances from natural sources or from industry and agriculture. Many of the naturally occurring contaminants in food are of microbiological origin and consist of harmful bacteria, bacterial toxins, and fungal toxins. (Aflatoxin, a contaminant of peanuts and grains, is an example of a fungal toxin or mycotoxin.) The second category of environmental contaminants includes organic chemicals, metals and their complexes, and radionuclides. Only those environmental contaminants introduced into food as a result of human activities such as agriculture, mining, and industry are considered in this assessment.

The environmental contamination of food is a result of our modern, high-technology society. We produce and consume large volumes of a wide variety of substances, some of which are toxic. It is estimated that 70,000 chemicals may currently be in commercial production in the United States and that 50 of these chemicals are manufactured in quantities greater than 1.3 billion lbs per year. Seven percent of this country's gross national product (GNP), $113 billion per year, is generated by the manufacture and distribution of chemicals (l). During the production, use, and disposal of these substances, there are opportunities for losses into the environment. For example, the Environmental Protection Agency (EPA) estimates that there are more than 30,000 chemical and radioactive waste disposal sites. Of these, 1,200 to 2,000 are considered threats to human health.

Environmental contamination of food takes two forms: long-term, low-level contamination resulting from gradual diffusion of persistent chemicals through the environment, and relatively shorter term, higher level contamination stemming from industrial accidents and waste disposal.

An example of low-level contamination is polychlorinated biphenyls (PCBs). This group of substances was widely used in transformers and capacitors, as heat-transfer fluids, and as an additive in dyes, carbon paper, pesticides, and plastics. Although production was halted in 1977, PCBs remain an ubiquitous, low-level contaminant of many foods, especially freshwater fish.

An example of the second type of contamination is polybrominated biphenyls (PBBs) in dairy products and meat. PBBs, a fire retardant, were accidentally mixed into animal feed. Dairy cattle that were fed the contaminated feed produced contaminated milk. The distinctions between the two types of food contamination are not exclusive. For exampie, PBBs have now become a long-term, low level contaminant in Michigan because they are very stable and resistant to decay. Animals raised on farms affected by the original feed contamination are now contaminated by the PBB residues remaining in the pastures and farm buildings.

How Food becomes Contaminated?

Chemicals contaminate foods through different routes depending on the chemical and its physical properties, its use, and the source or mechanism of contamination.

Organic substances that have contaminated food have been either industrial or agricultural chemicals. Pesticides are the only agricultural chemicals known to be environmental contaminants in food. A pesticide becomes an environmental contaminant when it is present in foods for which the application or use of the substance has not been approved. Livestock, poultry, and fish can be contaminated when application or manufacturing of pesticides occurs in the vicinity or when residues are transported through the environment. Improperly fumigated railroad cars, trucks, ships, or storage buildings used for transport or storage of human food and animal feed are also sources of environmental contamination. The interiors are sprayed or fumigated with pesticides, and if not sufficiently aired, contamination of the food or feed occurs.

The manufacture of organic chemicals produces sludges, gases, and liquid effluents of varying chemical complexities. The usual waste disposal methods (sewage systems, incineration, landfill) are unable to prevent organic residues from entering the environment in spite of Federal laws and corresponding regulations governing disposal. The routes include the atmosphere, soil, and surface or ground water.

Metals can be released into the environment in several ways. The mining and refining processes produce dust and gases which enter the atmosphere. Metallic salts formed during recovery and refining processes can escape as waste products into surface and ground water. Sewage sludge used as fertilizer on agricultural land also poses a potential food contamination problem. Trace metals present in the sludge can be taken up by crops grown on treated soil. Cadmium is the trace metal in sludge that currently generates the greatest concern.

Radioactivity in food stems from three sources: natural radioactivity. Releases from operation of nuclear reactors and processing plants, and fallout from nuclear weapons tests. The primary route by which food becomes contaminated is the deposition of airborne material on vegetation or soil. The subsequent fate of the radionuclide is determined by its chemical and physical nature and whether it is absorbed and metabolized by plants or animals. Natural radioactivity may become a concern when ores containing radioactive substances are mined and processed. The products or wastes may concentrate the radionuclides. Examples of this are uranium tailings, phosphate rock waste, or slags from phosphorus production. Radium may enter the food chain when it dissolves in ground water and is taken up through plant roots.

Nuclear reactors normally release radioactive noble gases that do not contaminate foods. Reactors do contain large inventories of fission products, transuranics, and other activation products. Accidental releases can contaminate vegetation by deposition of particles on leaves and soil, or through water. Gaseous releases would most likely involve the volatile elements such as iodine and tritium, or those with volatile precursors, such as strontium-90 and cesium-137. Aqueous releases would follow failure of the onsite ion exchange cleanup system. Any of the watersoluble elements could be involved.

Nuclear waste-processing plants could also have either gaseous or aqueous releases. In this case, the fission products are aged before processing, and iodine and the gaseous precursor radionuclides are not released. Tritium and carbon-14 are the major airborne products, while the waterborne radionuclides are the same as for reactors.

Atmospheric nuclear weapons tests distribute their fission products globally. Local deposition depends on the size of the weapon and the conditions of firing (high altitude, surface, or underground).

Magnitude of the Problem

There is little information available on the number of food contamination incidents, the amount and costs of food lost through regulatory actions, or the effects of consumption of contaminated food on health.

Evidence of Human Illness Resulting from Consumption of Contaminated Food

In evaluating the significance of environmental contaminants in food the key question is whether consumption of contaminated foods poses a health risk. Measurable health effects depend on the toxicity of the substance, the level at which it is present in food, the quantity of food consumed, and the vulnerability of the individual or population. In Japan, foods contaminated with substances such as PCBs, mercury, and cadmium have produced human illness and death. No such mass poisonings have occurred in the United States. However, in cases such as PBBs where a large populace has been exposed, some physiological changes have been noted. But no conclusions can as yet be drawn on the ultimate health effects.

It is known from limited surveys that the U.S. population is exposed to a wide variety of chemical contaminants through food, air, and water. The long-term health effects and the implications of possible interactions among these residues are unknown. A recent literature review of over 600 published studies found that nonoccupationally exposed

U.S. residents carry measurable residues of 94 chemical contaminants. Twentysix of these are organic substances, including twenty pesticides and pesticide metabolizes. The remainder are inorganic substances.

Americans also have been exposed to low levels of PCBs, PBBs, mercury, and ionizing radiation through their food. The following sections briefly summarize current knowledge and the extent of uncertainties on the health effects of these environmental contaminants.

Polychlorinated Biphenyls

PCBs occur in food as the result of environmental contamination leading to accumulation in the food chain, direct contact with food or animal feeds, or contact with food packaging materials made from recycled paper containing PCBs. Several comprehensive literature reviews have been published in the last 5 years detailing the acute and chronic toxic effects of PCBs in animals and humans.

Human illness has been caused by exposures to PCBs at much higher levels than those that occur in the United States. In the early part of 1968 the accidental contamination of edible rice-bran oil led to a poisoning epidemic among the Japanese families who consumed the oil. The disease became known as Yusho or rice-oil disease. Its chief symptoms were chloracne (a severe form of acne) and eye discharge; other symptoms included skin discoloration, headaches, fatigue, abdominal pain, menstrual changes, and liver disturbances. Babies born to mothers who consumed the rice oil were abnormally small and had temporary skin discoloration. The first symptoms of Yusho disease were registered on June 7, 1968, and 1,291 cases had been reported as of May 1975.

Since the rice oil was also contaminated with polychlorinated dibenzofuran (PCDF), it is difficult to determine from the Yusho data exactly what effect(s) exposure to PCBs alone could have on humans. It has been calculated that the PCDF made the rice oil 2 to 3.5 times more toxic than would have been expected from its PCB content alone. Careful records of the 1,291 Yusho patients have been kept to determine possible long-term effects. At least 9 of 29 deaths that occurred as of May 1975 were attributed to cancer (malignant neoplasm), but a causal relationship between PCBs and cancer cannot necessarily be inferred because of the high concentration of PCDF in the oil. The Yusho study, nevertheless, had two important results: first, the information established that PCBs can be transferred from mother to fetus and from mother to child through breast feeding, and second highly chlorinated PCB compounds are excreted more slowly from the body than less chlorinated ones.

More recent experiments in animals have demonstrated a variety of toxic effects. Cancers have been produced in mice and rats fed PCBs. Monkeys fed levels of PCBs equivalent to the amounts consumed by Yusho patients developed similar reproductive disorders. Young monkeys nursing on mothers consuming feed containing PCB developed toxic effects and behavioral abnormalities.

Polybrominated Biphenyls

Practically every Michigan resident has been exposed to PBB-contaminated food products. It is estimated that some 2,000 farm families who consumed products from their own PBB-contaminated farms have received the heaviest exposure.

Fries studied the kinetics of PBB absorption in dairy cattle and its elimination in milk, If intake of contaminated milk alone is considered, those Michigan residents most severely exposed consumed from 5 to 15 grams of PBB over the initial 230 days of the exposure. Those residents that coincidentally consumed contaminated meat and/or eggs may have received higher total doses of PBB, but the number of such cases is probably small.

Geographically the residents of the lower peninsula, where the original accident occurred, were found to have the greatest levels of exposure. In 1976, the Michigan Department of Public Health conducted a study on PBB concentrations in breast milk. It was found that 96 percent of the 53 women selected from the lower peninsula and 43 percent of the 42 women selected from the upper peninsula excreted PBB in their breast milk.

Low concentrations of PBBs also have been detected in animal feed in Indiana and Illinois. Unconfirmed surveys of food throughout the country found extremely low levels below the Food and Drug Administration (FDA) action level in the following states.

State Food	
Alabama	Chicken
Indiana	Turkey
Iowa	Beef
Mississippi	Chicken
New York	Chicken
Texas	Chicken
Wisconsin	Duck

Wolff, et al. reported that serum PBB was higher for males than females. It was suggested that the greater proportional body fat in women may account for this difference, but exposure may also be important. Males may consume more contaminated food or have more direct contact with PBB than females.

The same study found no consistent trends with respect to age. It was observed, however, that young males had greater concentrations of serum PBB than young females. Young females had greater concentrations than older males, and older males had greater concentrations than older females. It was also found that very young children and individuals who had lived on farms less than 1 year had lower serum PBB levels than other groups.

Serum PBB concentration is related to the intensity of exposure. Most studies indicate that consumers and residents of non-quarantined farms had significantly lower PBB levels than residents of quarantined farms; however, families on quarantined farms stopped consuming meat and milk from their own animals.

A sample of 165 exposed persons (quarantined farms) and 133 nonexposed (nonquarantined farms) was studied. Medical history interviews and physical examinations were performed on each subject and blood specimens were taken, Blood PBB levels as high as 2.26 parts per million (ppm) were found in the exposed individuals; about half exhibited levels greater than 0.02 ppm. Of the nonexposed individuals, only two showed blood PBB levels greater than 0.02 ppm; 70 percent of the adults and 97 percent of the children exhibited levels of 0.0002 to 0.019 ppm. Comparison of a list of selected conditions and complaints revealed no significant differences in the frequency of illness between the two groups. Physical examinations and clinical laboratory tests disclosed no effects attributable to "chronic" PBB exposure The effect of PBB exposure on white blood cell (lymphocyte) function of Michigan dairy farmers who consumed contaminated farm products was examined by Bekesi, et al. Forty-five members of Michigan farm families who had eaten PBB-contaminated food for periods of 3 months up to 4 years after the original accident were compared for immunological function to 46 Wisconsin farmers and 79 New York residents. All of the exposed individuals showed reduced lymphocyte function, and 40 percent showed abnormal production of lymphocytes. There were also significant increases in lymphocytes with no detectable surface markers ("null" cells). However, the short- and long-term health implications of these differences are not now known.

Lillis examined Michigan farmers and consumers of dairy products and found that the effect of PBB on humans was mainly neurological in nature. He found marked fatigue, hypersomnia, and decreased capacity for physical or mental work. Other symptoms included headache: dizziness: irritability; and musculoskeletal. Arthritis-like complaints-swelling of the joints with deformity, pain, and limitation of movement. Less severe gastrointestinal and dermatological complaints were also encountered.

Mercury and Methylmercury

Foods are the major source of human exposure to mercury. The mercury concentration in food is dependent on the type of food, the environmental level of mercury in the area where the food is produced, and the use of mercury-containing compounds in the agricultural and industrial production of the food. All living organisms have the ability to concentrate mercury, Therefore, all animal and vegetable tissues contain at least trace amounts. Several recent reviews have examined the health effects associated with consumption of mercury. The results of these reviews indicate that the effects of methylmercury poisoning become detectable in the most sensitive adults at blood levels of mercury of 20 to 50 μg/100 ml, hair levels from 50 to 120 mg/kg, and body burdens between 0.5 and 0.8 mg/kg body weight.

Since the Minamata Bay tragedy in Japan, the effects of chronic exposure to methylmercury have been well-documented. Mercury readily accumulates within the central nervous system, and clearance of mercury back into the bloodstream is slow. Consequently, the central nervous system is considered to be the critical target in chronic mercury exposure. The clinical symptoms of central nervous system involvement include headache, vertigo, vasomotor disturbance, ataxia, and pain and numbness in the extremities. The most prominent structural changes of the central nervous system resulting from chronic mercury exposure are diffuse cellular degeneration.

In evaluating the teratogenic hazards of mercury exposure to man, the placental transfer of mercury is particularly significant. Levels that are not toxic to pregnant women are sufficient to produce birth defects in their offspring. Transfer of methylmercury across the human placenta results in slightly higher blood levels in the infant at birth than in the mother. Table compares fetal and maternal blood concentrations in normally exposed populations in Japan, Sweden, and the United States.

In humans, the most widely reported fetal risk associated with maternal exposure to mercury is brain damage. The placental transfer of mercury and its effects on the human fetus were first recognized in the 1950's with the well-known outbreak of congenital Minamata disease in the towns of Minamata and Niigata, Japan. By 1959, 23 infants suffering from mental retardation and motor disturbances had been born to mothers exposed to methylmercury during their pregnancies. The clinical symptoms of the infants resembled those of severe cerebral palsy or cerebral dysfunction syndrome. They included disturbance of coordination, speech, and hearing; constriction of visual field; impairment of chewing and swallowing; enhanced tendon reflex; pathological reflexes; involuntary movement; primitive reflexes; superficial sensation; salivation; and forced laughing. Only 1 of the 23 mothers exhibited any symptoms of mercury poisoning.

Radioactivity

Ionizing radiation (X-rays, gamma rays, or beta particles with sufficient energy to strip electrons from molecules and produce ions) can produce birth defects, mutations, and cancers. These adverse health effects are usually associated with high dose levels delivered at high dose rates.

Such a combination is not ordinarily encountered in food. Previous radioactive contamination of foods has involved relatively small quantities of radioactive elements which have delivered low dose rates.

In these situations, the effects of the radia tion exposure on health are extremely difficult to evaluate. High dose rates (100 million to 1 billion times background) are estimated to produce 2,600 ionization events per second in cells. Background radiation levels are estimated to produce less than one ionization in the cell nucleus per day. Because cells have the capacity to repair damage to their genetic material, repair of ionization damage may occur at low radiation exposure. Higher exposures may overwhelm the cells' repair capacity. Whether any effects are observed in such cases depends on several factors. These include the dose delivered to the tissues, the nature of the emissions, and the metabolism of the cell. The following examples illustrate these points:

- Strontium-90 in food arouses most concern not only because of its long half-life but also because it behaves in the body in a manner somewhat similar to calcium. The replacement of bone calcium with strontium-go exposes tissues and cells covering the bone to radiation, In addition, bone marrow is subject to the ionizing radiation from the strontium-go. Thus, cancer of the bone-forming and bone-covering tissue as well as leukemias of the bone marrow blood-forming cells can possibly result.
- Iodine is concentrated by the thyroid gland. Radioiodines produced in atmospheric nuclear detonations or released from nuclear power stations are also taken up and concentrated by the thyroid, increasing the risk of thyroid cancer.
- Tritium, or radioactive hydrogen, combines chemically with oxygen to form water. Tritium derived from food would be widely distributed throughout the body exposing all tissues to radiation

The uncertainties surrounding the repair capacities of cells and the irreversible nature of the possible health effects have led to the adoption in the United States of a prudent policy toward low-level ionizing radiation. Since any amount of radiation is potentially harmful, unnecessary exposure should be avoided.

Economic Impact

The economic impact of an incident involving the environmental contamination of food includes the cost of condemned food, health costs, and the corresponding distributional effects and costs. The magnitude of the economic impact is determined by:

- The amount of food contaminated,
- The concentration of the contaminant in food,
- The chemical and toxicological characteristics of the contaminant, and
- The corresponding regulatory action taken on the contaminated food.

The initial regulatory action taken by Federal and State authorities may be the issuance of a warning or the establishment of either an action level or a tolerance. Action levels and tolerances establish a permissible level for the contaminant in food. Any food found to contain concentrations of the substances above this level is condemned and either destroyed or restricted from being marketed.

Costs of Food Condemned

In addition to the four factors listed above, the cost of condemned food is also affected by its position in the food production and marketing process at the time of condemnation, An action level or tolerance for a contaminant is the most important of the five factors. If no action level or tolerance is set, no food would be condemned and thus there would be no costs incurred. The impact of such a regulation will depend on the exact level of a substance that is allowed to be present in food.

The chemical properties of a contaminant are also important because of the potential for long-term effects on the amount of food affected. Since many contaminants biologically and chemically degrade slowly, their presence in the environment can mean food contamination above the action level or tolerance for many years after the source of the pollution has been stopped. The James River in Virginia, for example, is still closed to commercial fishing several years after kepone discharges into the river have been eliminated. The relative influence for each of these factors on the final cost will vary in each contamination incident.

Estimates of the cost of food condemned through regulatory action are most often expressed in dollars. Consequently, this cost is usually (and incorrectly) cited as representative of the total economic impact. Such costs were collected in OTA's State and Federal surveys. The data, however, only partially reflect the total economic impact for environmental contamination of food in the United States. This is because the cost of condemned food is only one component of the total economic impact of an incident. In addition, few of the incidents reported to OTA included data on the cost of food condemned. OTA estimates from the available data that the total cost of condemned food as a result of environmental contamination in the United States since 1968 is over $282 million. The only cost estimates used were those clearly stated for an incident by the reporting States or Federal agencies.

1. State Estimates: Of the 18 States reporting contamination incidents, only 6 provided data on the economic impact in dollar terms. Of those six, Michigan represents 99 percent of the total cost ($255 million) while reporting only 19 percent of the number of incidents in the 18 States. Indeed, Michigan accounts for 90 percent of the total costs reported in the United States while reporting only 7 percent of the incidents that occurred during the 1968-78 period. It must be recognized, however, that 84 percent of Michigan's costs are attributed to the PBB incident. Many incidents reported by State and Federal agencies are considerably smaller than the PBB episode. Thus, the PBB episode is an indication of how severe a contamination incident can be.

Some States reported the amount of food destroyed without estimating the cost. Kentucky, for example, reported the destruction of 400,000 lbs of milk since 1968 because of pesticide contamination. While such information can be converted into dollars, data on market position and price of product at time of confiscation are not readily available. Many States were unable to provide any estimates on either the cost or the amount of food condemned as a result of reported contamination incidents. New York (with PCBs) and Virginia (with kepone) are two States that could not provide cost estimates for food condemned as a result of environmental contamination. Virginia, however, has initiated a study to determine the economic impact of the kepone incident.

2. Federal Estimates: Of the two Federal agencies reporting information to OTA on environmental contaminant incidents, USDA's Food Safety and Quality Service (FSQS) reported food condemnation cost estimates. These estimates, however, only cover livestock and poultry—the food products over which FSQS has regulatory authority. FDA, which has regulatory authority over the remaining food commodities, did not estimate costs for reported environmental contamination incidents (70 percent of the Federal total). Thus, a significant proportion of the total costs for environmental contamination incidents requiring Federal action is unknown. Comparison of the two agency responses with the State responses reveals little duplication in the reporting of incidents.

FSQS cost estimates were determined by the number of animals or pounds destroyed multiplied by the market value at the time of confiscation. Since most of these animals were taken at the farm or wholesale level, the market value was the farm or wholesale price. Most of the losses resulting from FDA actions would be based on a wholesale or retail price because the seized products had advanced further in the marketing system. Therefore, their estimated costs would be greater than if they were seized at the production level (generally the case with FSQS seizures).

Summing up, the available data on the cost of condemned food is limited; consequently OTA's $282 million condemned-food estimate is likely to be a gross underestimation of the actual costs. The true cost would be impossible to estimate from this limited sample.

Health Costs

Health costs are also an important component of economic impact. These costs are incurred by the consumer whose health has or potentially can be affected adversely by a contaminant present in food. These adverse effects can cause illness and death, and the range of effects will vary depending on the toxicity of the contaminant, the concentration of the contaminant in food, and the amount of food consumed.

The concentration of contaminants has been at levels that have not produced immediate measurable and conclusive effects in exposed populations. Estimates are therefore made for the potential long term effects on exposed populations from various contaminants in food.

Health costs can be estimated from such projected health effects. Costs would include health care costs for treating illness and burial expenses associated with death. Additional costs would include estimated value of productive days or years lost from work due to the projected illness or death associated with the contaminant in food. All of these health-related costs, however, do not and cannot include the emotional and psychological impacts on those afflicted and their friends and families.

Distributional Effects and Costs

Distributional effects and costs involve the various people, groups, and organizations who are economically affected by an environmental contamination incident. Information on the extent and distribution of such effects and costs provides a clearer picture of the total economic impact on society. This information is usually couched in descriptive terms. Those who are economically affected are identified but the extent of the impact is seldom estimated in dollars. The exact distribution of costs from an incident through society is affected by the same five factors that influence the cost of condemned food.

Many of the distributional effects and costs for various types of environmental contaminant incidents are discussed in the following sections. The purpose of this discussion is not to identify all the distributional costs but rather to demonstrate the variety of effects and costs that can result from an incident.

1. Producers: Food producers are affected economically in different ways by contamination episodes. But all are affected directly when the food they produce is condemned. For example, food found contaminated at the farm level is confiscated and destroyed. This was the case for over 500 Michigan farmers whose dairy herds were partially or entirely destroyed. In such cases, farmers either replace their livestock, plant a new crop, or go out of business.

Farmers can be faced with severe economic hardship, since they are not always reimbursed financially for the animals or commodities confiscated. While insurance programs such as the Federal Crop Insurance Corporation are available to cover natural hazards which might destroy crops or livestock, such Federal assistance is not available to farmers for losses from environmental contamination. An injured farmer can obtain a loan at commercial rates or sue the responsible firm for compensation. But the loan and the interest add to a farmer's financial difficulties, and suing for compensation can take time that the farmer may not have.

The commercial fisher is faced with a different situation. If a river, lake, or species of fish is restricted because of environmental contamination, the fisher whose source of income depends on this species or waterway may have few employment alternatives.

The alternatives depend to some degree on the extent of the contamination. If the only waterway available in a section of a State or a whole State is closed to commercial fishing because of the contamination, the fisher's source of employment is eliminated until the restriction is ended. Since the restriction can last for years (depending on the chemical stability of the contaminant), the fisher either will have to move to other commercial fishing areas or seek other employment.

Food producers economically affected by the condemnation of contaminated food are likely to incur health costs. This is because many of the producers and their families regularly eat the food that they produce or harvest. Consequently they are exposed to the contaminated food at greater concentrations than the average consumer. This was the case for several farm families in Michigan.

2. Firms Held Accountable for Environmental Contamination: In most instances blame for a contamination incident can be established. Those accountable are subject to fines and lawsuits. Firms admitting responsibility often try to settle with producers out of court if possible. Most of the compensation is for the economic damages stemming from the destruction of food or loss of employment. Compensation for people whose health has been impaired as a result of eating contaminated food would be sought through civil litigation. Such litigation, however, is rare in this country, since the level of contamination in food is so low that demonstrating the necessary cause and effect is difficult.

Fines or compensation paid by the firms held accountable for the contamination are, in fact, poor indicators of the true costs incurred by the producers. This is because the settlement costs which are frequently negotiated or imposed bear little relationship to the actual costs incurred.

For example, compensation has been provided by Michigan Chemical Corporation and Farm Bureau Services, Inc., to many of the farmers whose livestock and poultry were destroyed following PBB contamination. Michigan Chemical and Farm Bureau Services have together paid more than $40 million in compensation from a jointly established insurance pool. In another case involving PCB-contaminated fish meal sold to poultry producers, Ralston Purina Company negotiated compensation for the 400,000 chickens destroyed. The cost of the compensation has not been disclosed.

3. Governments: Federal, State, and local governments also incur costs from an environmental contamination incident. Although the Federal Government and most State governments have agencies with programs to regulate or control food safety problems, these programs usually are not funded to handle the kind of long-term problems created by a PBB or kepone incident. The Michigan Department of Agriculture, for example, estimates it will spend $40 million to $60 million within the next 5 years to monitor and test for PBBs in animals and animal byproducts. This is money that could have been saved or spent for other programs if PBB contamination had not occurred. In order to recover its expenses from the PBB incident, the State of Michigan filed a lawsuit against both the Michigan Chemical Corporation and Farm Bureau Services, Inc.,

claiming more than $100 million in damage.

Federal involvement is limited unless the contaminated food is part of interstate commerce, Many of these incidents are not considered by the Federal Government to involve interstate commerce, FDA may provide technical assistance at the request of the Statc government when a contamination incident is regarded to be a local problem . These technical facilities and experts are available to all States through the Federal and regional offices. Additional expenditures by the Federal Government for contamination incidents are limited. Additional State expenditures, however, can be substantial. Federal expenditures are made when Federal regulations are developed and promulgated for particular contaminants in food such as PCB.

4. Consumers: Consumers can incur costs from an environmental contaminant incident in several ways. The removal of food from commerce could increase prices for that food product or other food products being sold. Thus, the consumer could pay more for food as a result of an environmental contamination incident. In order for this price increase in food to occur, however, a significant amount of a food product or food products would have to be taken off the market. Such prices of food might vary by State or region and affect certain socioeconomic classes differently.

Health costs could increase as a result of the consumption of contaminated food. This would not affect all consumers but rather those who received the most exposure and/or those most susceptible to a contaminant, such as children or senior citizens. While these costs would already be included in estimated total health costs, the distributional effects could indicate those consumers most likely to be affected.

5. Indirect Costs: Most of the costs mentioned directly stem from an environmental contamination incident. However, indirect or secondary costs can and do occur. For example, a bait and tackle store on a lake that is closed to commercial and sport fishing because of an environmental contamination is likely to suffer economic hardship, Food processors whose normal supply of food has been condemned because of environmental contamination will also suffer economically unless they find new sources of supply. These are just two examples of the many indirect costs which might occur.

Potential Food Contamination Problems

Because a limited number of substances posing health problems already have been identified in food, concern exists that other toxic substances are likely to contaminate food in the future. This concern arises from the number of substances presently being manufactured, used, and disposed of in the United States, and the difficulties in preventing them from entering the environment. New substances developed to meet new needs or to replace known toxic substances may create unexpected environmental problems if not properly controlled. Byproducts of new technologies such as synthetic fuels are also potential environmental contaminants.

There are two methods of objectively assessing possible future contaminants: 1) by sampling the food supply for chemical contaminants and ranking them according to potential hazard and 2) by surveying the universe of industrial chemicals and ranking them according to their potential for entering the food supply in toxic amounts.

Of the three categories of environmental contaminants considered in this report, organic chemicals probably pose the greatest potential environmental and food contamination problems. This conclusion is based on the number, volume, and toxicity of the organics manufactured and used in this country. Both trace metals and radioactive substances continue to warrant concern, but not as great a concern as organic substances. The extent of food contamination from these substances depends on our success in preventing them from entering the environment.

BENZENE IN SOFT DRINKS

Benzene in soft drinks is of potential concern due to the carcinogenic nature of the benzene molecule. This contamination is a public health concern and has caused significant outcry among environmental and health advocates. Benzene levels are regulated in drinking water nationally and internationally, and in bottled water in the United States, but only informally in soft drinks. The benzene forms from decarboxylation of the preservative benzoic acid in the presence of ascorbic acid (vitamin C) and metal ions (iron and copper) that act as catalysts, especially under heat and light.

Formation in Soft Drinks

The major cause of benzene in soft drinks is the decarboxylation of benzoic acid in the presence of ascorbic acid (vitamin C, E300) or erythorbic acid (a diastereomer of ascorbic acid, E315). Benzoic acid is often added to drinks as a preservative in the form of its salts sodium benzoate (E211), potassium benzoate (E 212), or calcium benzoate (E 213). Citric acid is not thought to induce significant benzene production in combination with benzoic acid, but some evidence suggests that in the presence of ascorbic or erythorbic acid and benzoic acid, citric acid may accelerate the production of benzene.

The proposed mechanism begins with hydrogen abstraction by the hydroxyl radical, which itself is produced by the Cu^{2+}-catalysed reduction of dioxygen by ascorbic acid:

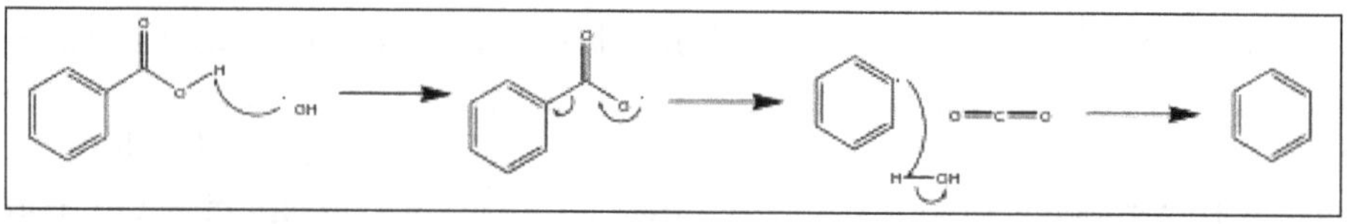

Other factors that affect the formation of benzene are heat and light. Storing soft drinks in warm conditions speeds up the formation of benzene.

Calcium disodium EDTA and sugars have been shown to inhibit benzene production in soft drinks.

The International Council of Beverages Associations (ICBA) has produced advice to prevent or minimize benzene formation.

Environmental Exposure to Benzene

Benzene in soft drinks has to be seen in the context of other environmental exposure. Taking the worst example found to date of a soft drink containing 87.9 ppb benzene, someone drinking a 350 ml (12 oz) can would ingest 31 µg (micrograms) of benzene, almost equivalent to the benzene inhaled by a motorist refilling a fuel tank for three minutes. While there are alternatives to using sodium benzoate as a preservative, the casual consumption of such a drink is unlikely to pose a significant health hazard to a particular individual.

The UK Food Standards Agency has stated that people would need to drink at least 20 litres (5.5 gal) per day of a drink containing benzene at 10 µg to equal the amount of benzene they would breathe from city air every day. Daily personal exposure to benzene is determined by adding exposure from all sources.

- Air: A European study found that people breathe in 220 µg of benzene every day due to general atmospheric pollution. A motorist refilling a fuel tank for three minutes would inhale a further 32 µg. The estimated daily exposure from "automobile-related activities" is 49 µg and for driving for one hour is 40 µg.
- Smoking: For smokers, cigarette smoking is the main source of exposure to benzene. Estimates are 7900 µg per day (smoking 20 cigarettes per day), 1820 µg/day, and 1800 µg/day.
- Passive smoking: Benzene intake from passive smoking is estimated at 63 µg/day (Canada) and 50 µg/day.
- Diet and drinking water: 0.2 to 3.1 µg per day.

FOOD MOLDS

Molds are microscopic fungi that live on plant or animal matter. Mold grows from tiny spores that float around in the air. When some of these spores fall onto a piece of damp food, they grow into mold.

Food mold feeds itself by producing chemicals that make the food break down and start to rot. As the bread rots, the mold grows. There are thousands of different kinds of molds. One mold that grows on lemons looks like a blue-green powder. A mold that grows on strawberries is a grayish-white fuzz. A common mold that grows on bread looks like white cottony fuzz at first. If you watch that mold for a few days, it will turn black. The tiny black dots are its spores, which can grow to produce more mold.

No one knows how many species of fungi exist, but estimates range from tens of thousands to perhaps 300,000 or more. Most are filamentous (thread like) organisms and the production of spores is characteristic of fungi in general. These spores can be transported by air, water, or insects.

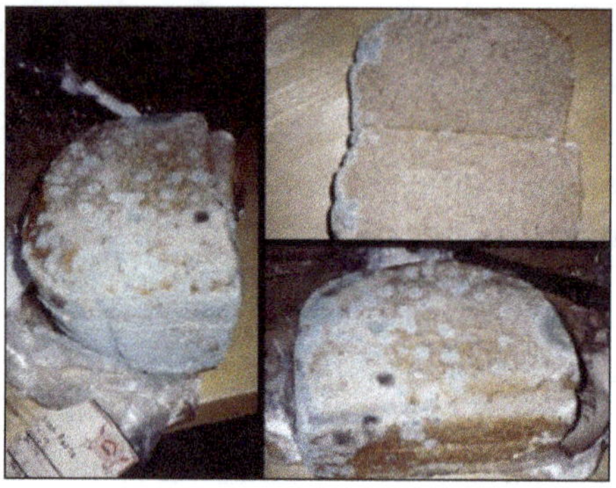

Mold on Bread.

Mold on Tomatoes.

Are Molds Dangerous?

Yes, some molds cause allergic reactions and respiratory problems. And a few molds, in the right conditions, produce mycotoxins, poisonous substances that can make you sick.

- Mycotoxins: Mycotoxins are poisonous substances produced by certain molds found primarily in grain and nut crops, but are also known to be on celery, grape juice, apples, and other produce. There are many of them and scientists are continually discovering new ones. The Food and Agriculture Organization (FAO) of the United Nations estimates that 25% of the world's food crops are affected by mycotoxins, of which the most notorious are aflatoxins.
- Aflatoxin: Aflatoxin is a cancer-causing poison produced by certain fungi in or on foods and feeds, especially in field corn and peanuts. They are probably the best known and most intensively researched mycotoxins in the world. Aflatoxins have been associated with various diseases, such as aflatoxicosis in livestock, domestic animals, and humans throughout the world. Many countries try to limit exposure to aflatoxin by regulating and monitoring its presence on commodities intended for use as food and feed. The prevention of aflatoxin is one of the most challenging toxicology issues of present time.
- Cheese Molds: An exception is mold on hard cheese, as some cheeses are eaten only after they become moldy Blue cheese gets its flavor from the veins of blue-green mold in it. When a blue cheese is formed into a wheel, holes are

poked through it with thin skewers. Air gets into these holes, and a very special kind of mold grows there as the cheese ripens. If mold develops, cut away one inch on each side of the cheese (throw away) and use the remainder as soon as possible.

Some moldy cheeses are safe to eat after the mold has been sliced off, while others are toxic.

Hard and semisoft cheese, such as parmesan, Swiss, romano and cheddar, you can cut away the moldy part and eat the rest of the cheese. Cut off at least 1-inch around and below the moldy spot.

With soft cheeses, such as brie, chevre (goat cheese), blue cheese, and ricotta, the mold that grows on these cheeses cannot be safely removed so they should be discarded. One reason is that the molds can more easily penetrate into the heart of soft cheeses than they can into harder cheeses. This causes spoilage from within that cannot be scraped away. The same goes for any cheese that has been shredded, crumbled or sliced. If mold is found on soft cheese (i.e. cottage cheese, cream cheese) the entire package should be discarded. Mold on soft cheeses are toxic.

You only see part of the mold on the surface of food – gray fur on forgotten bologna, fuzzy green dots on bread, white dust on Cheddar, coin-size velvety circles on fruits, and furry growth on the surface of jellies. When a food shows heavy mold growth, "root" threads have invaded it deeply. In dangerous molds, poisonous substances are often contained in and around these threads. In some cases, toxins may have spread throughout the food.

Why can Mold Grow in the Refrigerator?

While most molds prefer warmer temperatures, they can grow at refrigerator temperatures, too. Molds also tolerate salt and sugar better than most other food invaders. Therefore, molds can grow in refrigerated jams and jelly and on cured, salty meats (ham, bacon, salami, and bologna).

- Cleanliness is vital in controlling mold, because mold spores from contaminated food can build up in your refrigerator, dishcloths and other cleaning utensils.
- Clean the refrigerator or pantry at the spot where the food was stored. Check nearby items the moldy food might have touched. Mold spreads quickly in fruits and vegetables.
- Clean the inside of the refrigerator every few months with 1 tablespoon of baking soda dissolved in a quart of water. Rinse with clear water and dry. Scrub visible mold (usually black) on rubber casings using 3 teaspoons of bleach in a quart of water.

- Keep dishcloths, towels, sponges and mops clean and fresh. A musty smell means they're spreading mold around. Discard items you can not clean or launder.
- Keep the humidity level in the house as low as practical – below 40 percent, if possible.

How can you Protect Food from Mold?

When serving food, keep it covered to prevent exposure to mold spores in the air. Use plastic wrap to cover foods you want to stay moist (fresh or cut fruits and vegetables, and green and mixed salads).

- Empty opened cans of perishable foods into clean storage containers and refrigerate them promptly.
- Don't leave any perishables out of the refrigerator more than 2 hours.
- Use leftovers within 3 to 4 days so mold does not have a chance to grow.

How should you Handle Food with Mold on it?

Buying small amounts and using food quickly can help prevent mold growth. But when you see moldy food:

- Do not sniff the moldy item. This can cause respiratory trouble.
- If food is covered with mold, discard it. Put it into a small paper bag or wrap it in plastic and dispose in a covered trash can that children and animals can not get into.
- Clean the refrigerator or pantry at the spot where the food was stored.
- Check nearby items the moldy food might have touched. Mold spreads quickly in fruits and vegetables.

Molds on Food

- Luncheon Meats, Bacon, or Hot Dogs: Foods with high moisture content can be contaminated below the surface. Moldy foods may also have bacteria growing along with the mold.
- Hard Salami and Dry-cured Country Hams: Scrub mold off surface. It is normal for these shelf-stable products to have surface mold.
- Cooked Leftover Meat and Poultry: Foods with high moisture content can be contaminated below the surface. Moldy foods may also have bacteria growing along with the mold.

- Cooked Casseroles: Foods with high moisture content can be contaminated below the surface. Moldy foods may also have bacteria growing along with the mold.
- Cooked Grain and Pasta: Foods with high moisture content can be contaminated below the surface. Moldy foods may also have bacteria growing along with the mold.
- Hard Cheese (not Cheese where Mold is Part of the Processing): Cut off at least 1 inch around and below the mold spot (keep the knife out of the mold itself so it will not cross-contaminate other parts of the cheese). After trimming off the mold, re-cover the cheese in fresh wrap. Mold generally cannot penetrate deep into the product.

  Cheese made with mold (such as Roquefort, blue, Gorgonzola, Stilton, Brie, Camembert) Soft cheeses such as Brie and Camembert if they contain molds that are not a part of the manufacturing process. If surface mold is on hard cheeses such as Gorgonzola and Stilton, cut off mold at least 1 inch around and below the mold spot and handle like hard cheese (above). Molds that are not a part of the manufacturing process can be dangerous.

  Soft cheese (such as cottage, cream cheese, Neufchatel, chevre, Bel Paese, etc.) Foods with high moisture content can be contaminated below the surface. Shredded, sliced, or crumbled cheese can be contaminated by the cutting instrument. Moldy soft cheese can also have bacteria growing along with the mold.
- Yogurt and Sour Cream: Foods with high moisture content can be contaminated below the surface. Moldy foods may also have bacteria growing along with the mold.
- Jams and Jellies: The mold could be producing a mycotoxin. Microbiologists recommend against scooping out the mold and using the remaining condiment.
- Fruits and Vegetables, Firm (Such as Cabbage, Bell Peppers, Carrots, etc.): Cut off at least 1 inch around and below the mold spot (keep the knife out of the mold itself so it will not cross-contaminate other parts of the produce). Small mold spots can be cut off fruits and vegetables with low moisture content. It is difficult for mold to penetrate dense foods.

  Fruits and vegetables, soft (such as cucumbers, peaches, tomatoes, etc.)

  Fruits and vegetables with high moisture content can be contaminated below the surface.
- Bread and Baked Goods: Porous foods can be contaminated below the surface.
- Peanut Butter, Legumes and Nuts: Foods processed without preservatives are at high risk for mold.

CROSS CONTAMINATION

Cross contamination occurs when bacteria and viruses are transferred from a contaminated food or surface such as a chopping board and utensils to other food. For example, it can happen when bacteria from the surface of raw meat, poultry, seafood and raw vegetables (such as unwashed potatoes and other root vegetables), are transferred onto ready to eat foods, such as leaf and vegetable salads, rice or pasta salads, cooked meats, poultry, seafood or even fruit. The bacteria on the raw food are killed when the food is cooked, but the ready to eat food is eaten without further cooking – bacteria, viruses and all.

Hands are among the obvious culprits in transferring bacteria and viruses from raw to ready to eat food, but direct contact with soiled raw foods, dirty chopping boards, knives and other food preparation implements and containers can also spread the contamination. Chopping boards, plates and knives, blenders, mixers, bowls, or any other surface that has been in contact with raw meats, seafood and soiled vegetables and herbs needs to be carefully washed with warm water and detergent, then rinsed and thoroughly dried before being used for ready to eat foods.

Cross contamination can also occur from incorrectly storing raw food in the fridge. If raw food is placed in direct contact with ready to eat foods, or if raw meat juices drip onto cooked foods, fruit and other ready to eat food, cross contamination can occur.

Raw foods should always be treated as though they are contaminated. Raw food, such as meat, poultry or seafood, should be stored in a rigid leak proof container or at the bottom of the fridge to prevent it coming into direct contact with ready to eat food or to prevent meat juices or liquids dripping onto other food. Ready to eat food should be stored covered in the fridge to further reduce the risks.

In the home it really doesn't matter whether you have wooden, plastic or glass chopping boards so long as they are kept really clean and in good condition. The porous nature of wood makes it advisable to use plastic or glass chopping boards for raw meat, poultry and seafood. It is easier if you have two boards – one used only for raw food and one for cooked and ready to eat food or bread – to prevent cross contamination. All chopping boards should be scrubbed with hot water and detergent and dried after preparing raw foods. Plastic chopping boards are good as they can be washed at high water temperatures in the dishwasher. All plastic and wooden cutting boards wear out over time. Once cutting boards become excessively worn or develop hard-to-clean grooves, they should be discarded.

References

- Food-safety-and-the-different-types-of-food-contamination: foodsafety.com.au, Retrieved 15 January, 2020
- What-are-physical-contaminants-in-food-manufacturing: thermofisher.com, Retrieved 12 July, 2020

- Investigation-and-identification-of-physical-contaminants-in-food: foodsafetymagazine.com, Retrieved 22 June, 2020
- Adulteration, health-campaigns/beware-of-adulteration: vikaspedia.in, Retrieved 02 January, 2020
- Modern-analysis-of-chemical-contaminants-in-food: foodsafetymagazine.com, Retrieved 25 May, 2020
- Mold: whatscookingamerica.net, Retrieved 05 August, 2020
- Cross-contamination: foodsafety.asn.au, Retrieved 26 January, 2020

3

Food Preservation

The food can be preserved by using many preservation techniques such as, pickling, canning, pasteurization, food drying, biopreservation, fermentation, chemical preservation, low-temperature preservation, thermal and non-thermal processing of food, etc. This chapter discusses these varied food preservation techniques in detail.

Food preservation refers to any one of a number of techniques used to prevent food from spoiling. It includes methods such as canning, pickling, drying and freeze-drying, irradiation, pasteurization, smoking, and the addition of chemical additives. Food preservation has become an increasingly important component of the food industry as fewer people eat foods produced on their own lands, and as consumers expect to be able to purchase and consume foods that are out of season.

The vast majority of instances of food spoilage can be attributed to one of two major causes: (1) the attack by pathogens (disease-causing microorganisms) such as bacteria and molds, or (2) oxidation that causes the destruction of essential biochemical compounds and/or the destruction of plant and animal cells. The various methods that have been devised for preserving foods are all designed to reduce or eliminate one or the other (or both) of these causative agents.

For example, a simple and common method of preserving food is by heating it to some minimum temperature. This process prevents or retards spoilage because high temperatures kill or inactivate most kinds of pathogens. The addition of compounds known as BHA and BHT to foods also prevents spoilage in another different way. These compounds are known to act as antioxidants, preventing chemical reactions that cause the oxidation of food that results in its spoilage. Almost all techniques of preservation are designed to extend the life of food by acting in one of these two ways.

The search for methods of food preservation probably can be traced to the dawn of human civilization. People who lived through harsh winters found it necessary to find some means of insuring a food supply during seasons when no fresh fruits and vegetables were available. Evidence for the use of dehydration (drying) as a method of food preservation, for example, goes back at least 5,000 years. Among the most primitive forms of food preservation that are still in use today are such methods as smoking, drying, salting, freezing, and fermenting.

Early humans probably discovered by accident that certain foods exposed to smoke seem to last longer than those that are not. Meats, fish, fowl, and cheese were among such foods. It appears that compounds present in wood smoke have anti-microbial actions that prevent the growth of organisms that cause spoilage. today, the process of smoking has become a sophisticated method of food preservation with both hot and cold forms in use. Hot smoking is used primarily with fresh or frozen foods, while cold smoking is used most often with salted products. The most advantageous conditions for each kind of smoking—air velocity, relative humidity, length of exposure, and salt content, for example—are now generally understood and applied during the smoking process. For example, electrostatic precipitators can be employed to attract smoke particles and improve the penetration of the particles into meat or fish. So many alternative forms of preservation are now available that smoking no longer holds the position of importance it once did with ancient peoples. More frequently, the process is used to add interesting and distinctive flavors to foods.

Because most disease-causing organisms require a moist environment in which to survive and multiply, drying is a natural technique for preventing spoilage. Indeed, the act of simply leaving foods out in the sun and wind to dry out is probably one of the earliest forms of food preservation. Evidence for the drying of meats, fish, fruits, and vegetables go back to the earliest recorded human history. At some point, humans also learned that the drying process could be hastened and improved by various mechanical techniques. For example, the Arabs learned early on that apricots could be preserved almost indefinitely by macerating them, boiling them, and then leaving them to dry on broad sheets. The product of this technique, quamar-adeen, is still made by the same process in modern Muslim countries.

Today, a host of dehydrating techniques are known and used. The specific technique adopted depends on the properties of the food being preserved. For example, a traditional method for preserving rice is to allow it to dry naturally in the fields or on drying racks in barns for about two weeks. After this period of time, the native rice is threshed and then dried again by allowing it to sit on straw mats in the sun for about three days. Modern drying techniques make use of fans and heaters in controlled environments. Such methods avoid the uncertainties that arise from leaving crops in the field to dry under natural conditions. Controlled temperature air drying is especially popular for the preservation of grains such as maize, barley, and bulgur.

Vacuum drying is a form of preservation in which a food is placed in a large container from which air is removed. Water vapor pressure within the food is greater than that outside of it, and water evaporates more quickly from the food than in a normal atmosphere. Vacuum drying is biologically desirable since some enzymes that cause oxidation of foods become active during normal air drying. These enzymes do not appear to be as active under vacuum drying conditions, however. Two of the special advantages of vacuum drying are that the process is more efficient at removing water from a food product, and it takes place more quickly than air drying. In one study, for example, the drying time of a fish fillet was reduced from about 16 hours by air drying to six hours as a result of vacuum drying.

Coffee drinkers are familiar with the process of dehydration known as spray drying. In this process, a concentrated solution of coffee in water is sprayed though a disk with many small holes in it. The surface area of the original coffee grounds is increased many times, making dehydration of the dry product much more efficient. Freeze-drying is a method of preservation that makes use of the physical principle known as sublimation. Sublimation is the process by which a solid passes directly to the gaseous phase without first melting. Freeze-drying is a desirable way of preserving food because at low temperatures (commonly around 14°F to −13°F [−10°C to −25°C]) chemical reactions take place very slowly and pathogens have difficulty surviving. The food to be preserved by this method is first frozen and then placed into a vacuum chamber. Water in the food first freezes and then sublimes, leaving a moisture content in the final product of as low as 0.5%.

The precise mechanism by which salting preserves food is not entirely understood. It is known that salt binds with water molecules and thus acts as a dehydrating agent in foods. A high level of salinity may also impair the conditions under which pathogens can survive. In any case, the value of adding salt to foods for preservation has been well known for centuries. Sugar appears to have effects similar to those of salt in preventing spoilage of food. The use of either compound (and of certain other natural materials) is known as curing. A desirable side effect of using salt or sugar as a food preservative is, of course, the pleasant flavor each compound adds to the final product.

Curing can be accomplished in a variety of ways. Meats can be submerged in a salt solution known as brine, for example, or the salt can be rubbed on the meat by hand. The injection of salt solutions into meats has also become popular. Food scientists have now learned that a number of factors relating to the food product and to the preservative conditions affect the efficiency of curing. Some of the food factors include the type of food being preserved, the fat content, and the size of treated pieces. Preservative factors include brine temperature and concentration, and the presence of impurities.

Curing is used with certain fruits and vegetables, such as cabbage (in the making of sauerkraut), cucumbers (in the making of pickles), and olives. It is probably most popular, however, in the preservation of meats and fish. Honey-cured hams, bacon, and corned beef ("corn" is a term for a form of salt crystals) are common examples.

Freezing is an effective form of food preservation because the pathogens that cause food spoilage are killed or do not grow very rapidly at reduced temperatures. The process is less effective in food preservation than are thermal techniques such as boiling because pathogens are more likely to be able to survive cold temperatures than hot temperatures. In fact, one of the problems surrounding the use of freezing as a method of food preservation is the danger that pathogens deactivated (but not killed) by the process will once again become active when the frozen food thaws.

A number of factors are involved in the selection of the best approach to the freezing of foods, including the temperature to be used, the rate at which freezing is to take place, and the actual method used to freeze the food. Because of differences in cellular composition, foods actually begin to freeze at different temperatures ranging from about 31 °F (–0.6 °C) for some kinds of fish to 19 °F (–7 °C) for some kinds of fruits.

The rate at which food is frozen is also a factor, primarily because of aesthetic reasons. The more slowly food is frozen, the larger the ice crystals that are formed. Large ice crystals have the tendency to cause rupture of cells and the destruction of texture in meats, fish, vegetables, and fruits. In order to deal with this problem, the technique of quick-freezing has been developed. In quick-freezing, a food is cooled to or below its freezing point as quickly as possible. The product thus obtained, when thawed, tends to have a firm, more natural texture than is the case with most slow-frozen foods.

About a half dozen methods for the freezing of foods have been developed. One, described as the plate, or contact, freezing technique, was invented by the American inventor Charles Birdseye in 1929. In this method, food to be frozen is placed on a refrigerated plate and cooled to a temperature less than its freezing point. Alternatively, the food may be placed between two parallel refrigerated plates and frozen. Another technique for freezing foods is by immersion in very cold liquids. At one time, sodium chloride brine solutions were widely used for this purpose. A 10% brine solution, for example, has a freezing point of about 21 °F (–6 °C), well within the desired freezing range for many foods. More recently, liquid nitrogen has been used for immersion freezing. The temperature of liquid nitrogen is about –320 °F (–195.5 °C), so that foods immersed in this substance freeze very quickly.

As with most methods of food preservation, freezing works better with some foods than with others. Fish, meat, poultry, and citrus fruit juices (such as frozen orange juice concentrate) are among the foods most commonly preserved by this method.

Fermentation is a naturally occurring chemical reaction by which a natural food is converted into another form by pathogens. It is a process in which food spoils, but results in the formation of an edible product. Perhaps the best example of such a food is cheese. Fresh milk does not remain in edible condition for a very long period of time. Its pH is such that harmful pathogens begin to grow in it very rapidly. Early humans discovered, however, that the spoilage of milk can be controlled in such a way as to produce a new product, cheese.

Bread is another food product made by the process of fermentation. Flour, water, sugar, milk, and other raw materials are mixed together with yeasts and then baked. The addition of yeasts brings about the fermentation of sugars present in the mixture, resulting in the formation of a product that will remain edible much longer than will the original raw materials used in the bread-making process.

Heating food is an effective way of preserving it because the great majority of harmful pathogens are killed at temperatures close to the boiling point of water. In this respect, heating foods is a form of food preservation comparable to that of freezing but much superior to it in its effectiveness. A preliminary step in many other forms of food preservation, especially forms that make use of packaging, is to heat the foods to temperatures sufficiently high to destroy pathogens.

In many cases, foods are actually cooked prior to their being packaged and stored. In other cases, cooking is neither appropriate nor necessary. The most familiar example of the latter situation is pasteurization. During the 1860s, the French bacteriologist Louis Pasteur discovered that pathogens in foods could be destroyed by heating those foods to a certain minimum temperature. The process was particularly appealing for the preservation of milk since preserving milk by boiling is not a practical approach. Conventional methods of pasteurization called for the heating of milk to a temperature between 145 and 149 °F (63-65 °C) for a period of about 30 minutes, and then cooling it to room temperature. In a more recent revision of that process, milk can also be "flash-pasteurized" by raising its temperature to about 160 °F (71 °C) for a minimum of 15 seconds, with equally successful results. A process known as ultra-high-pasteurization uses even higher temperatures, of the order of 194-266 °F (90-130 °C), for periods of a second or more.

One of the most common methods for preserving foods today is to enclose them in a sterile container. The term "canning" refers to this method although the specific container can be glass, plastic, or some other material as well as a metal can, from which the procedure originally obtained its name. The basic principle behind canning is that a food is sterilized, usually by heating, and then placed within an airtight container. In the absence of air, no new pathogens can gain access to the sterilized food. In most canning operations, the food to be packaged is first prepared—cleaned, peeled, sliced, chopped, or treated in some other way—and then placed directly into the container. The container is then placed in hot water or some other environment where its temperature is raised above the boiling point of water for some period of time. This heating process achieves two goals at once. First, it kills the vast majority of pathogens that may be present in the container. Second, it forces out most of the air above the food in the container.

After heating has been completed, the top of the container is sealed. In home canning procedures, one way of sealing the (usually glass) container is to place a layer of melted paraffin directly on top of the food. As the paraffin cools, it forms a tight solid seal on top of the food. Instead of or in addition to the paraffin seal, the container is also sealed with a metal screw top containing a rubber gasket. The first glass jar designed for this type of home canning operation, the Mason jar, was patented in 1858.

The commercial packaging of foods frequently makes use of tin, aluminum, or other kinds of metallic cans. The technology for this kind of canning was first developed in the mid-1800s, when individual workers hand-sealed cans after foods had been cooked within them. At this stage, a single worker could seldom produce more than 100 "canisters" (from which the word "can" later came) of food a day. With the development of far more efficient canning machines in the late nineteenth century, the mass production of canned foods became a reality.

As with home canning, the process of preserving foods in metal cans is simple in concept. The foods are prepared and the empty cans are sterilized. The prepared foods are then added to the sterile metal can, the filled can is heated to a sterilizing temperature, and the cans are then sealed by a machine.

The majority of food preservation operations used today also employ some kind of chemical additive to reduce spoilage. Of the many dozens of chemical additives available, all are designed either to kill or retard the growth of pathogens or to prevent or retard chemical reactions that result in the oxidation of foods. Some familiar examples of the former class of food additives are sodium benzoate and benzoic acid; calcium, sodium propionate, and propionic acid; calcium, potassium, sodium sorbate, and sorbic acid; and sodium and potassium sulfite. Examples of the latter class of additives include calcium, sodium ascorbate, and ascorbic acid (vitamin C); butylated hydroxyanisole (BHA) and buty-lated hydroxytoluene (BHT); lecithin; and sodium and potassium sulfite and sulfur dioxide.

A special class of additives that reduce oxidation is known as the sequestrants. Sequestrants are compounds that "capture" metallic ions, such as those of copper, iron, and nickel, and remove them from contact with foods. The removal of these ions helps preserve foods because in their free state they increase the rate at which oxidation of foods takes place. Some examples of sequestrants used as food preservatives are ethylenediamine-tetraacetic acid (EDTA), citric acid, sorbitol, and tartaric acid.

CURING

Curing is any of various food preservation and flavoring processes of foods such as meat, fish and vegetables, by the addition of salt, with the aim of drawing moisture out of the food by the process of osmosis. Because curing increases the solute concentration in the food and hence decreases its water potential, the food becomes inhospitable for the microbe growth that causes food spoilage. Curing can be traced back to antiquity, and was the primary way of preserving meat and fish until the late-19th century. Dehydration was the earliest form of food curing. Many curing processes also involve smoking, spicing, cooking, or the addition of combinations of sugar, nitrate, nitrite.

Meat preservation in general (of meat from livestock, game, and poultry) comprises the set of all treatment processes for preserving the properties, taste, texture, and color of raw, partially cooked, or cooked meats while keeping them edible and safe to consume.

Curing has been the dominant method of meat preservation for thousands of years, although modern developments like refrigeration and synthetic preservatives have begun to complement and supplant it.

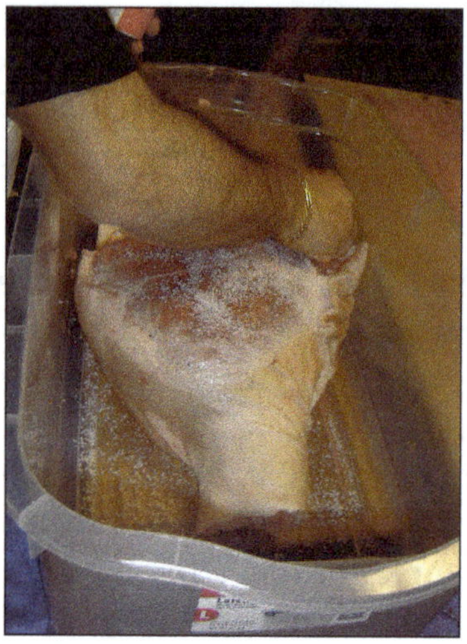

Sea salt being added to raw ham to make prosciutto.

Slices of beef in a can.

While meat-preservation processes like curing were mainly developed in order to prevent disease and to increase food security, the advent of modern preservation methods mean that in most developed countries today curing is instead mainly practised for its cultural value and desirable impact on the texture and taste of food. For lesser-developed countries, curing remains a key process in the production, transport and availability of meat.

Curing salt, also known as "Prague powder" or "pink salt", is typically a combination of sodium chloride and sodium nitrite that is dyed pink to distinguish it from table salt.

Some traditional cured meat (such as authentic Parma ham and some authentic Spanish chorizo and Italian salami, etc.) are cured with salt alone. Today, potassium nitrate and sodium nitrite (in conjunction with salt) are the most common agents in curing meat, because they bond to the myoglobin and act as a substitute for the oxygen, thus turning myoglobin red . More recent evidence shows that these chemicals also inhibit the growth of the bacteria that cause the disease botulism. The combination of table salt with nitrates or nitrites, called curing salt, is often dyed pink to distinguish it from table salt. Neither table salt, nor any of the nitrites or nitrates commonly used in curing (e.g. sodium nitrate, sodium nitrite, and potassium nitrate) is naturally pink.

Necessity of Curing

Untreated meat decomposes rapidly if it is not preserved, at a speed that depends on several factors, including ambient humidity, temperature, and the presence of pathogens. Most meats cannot be kept at room temperature in excess of a few days without spoiling.

If kept in excess of this time, meat begins to change color and exude a foul odor, indicating the decomposition of the food. Ingestion of such spoiled meat can cause serious food poisonings, like botulism. Salt-curing processes have been developed since antiquity in order to ensure food safety without relying on artificial anti-bacterial agents.

While the short shelf life of fresh meat does not pose a significant problem when access to it is easy and supply is abundant, in times of scarcity and famine, or when the meat must be carried over long voyages, it spoils very quickly. In such circumstances the usefulness of preserving foods containing nutritional value for transport and storage is obvious.

Curing can significantly extend the life of meat before it spoils, by making it inhospitable to the growth of spoilage microbes.

A survival technique since prehistory, the preservation of meat has become, over the centuries, a topic of political, economic, and social importance worldwide.

Traditional Methods

Food curing dates back to ancient times, both in the form of smoked meat and salt-cured meat. Several sources describe the salting of meat in the ancient Mediterranean world. Diodore of Sicily in his *Bibliotheca historica* wrote that the Cosséens in the mountains of Persia salted the flesh of carnivorous animals. Strabo indicates that people at Borsippa were catching bats and salting them to eat. The ancient Greeks prepared *tarichos*, which was meat and fish conserved by salt or other means. The Romans called this dish *salsamentum* – which term later included salted fat, the sauces and spices used for its preparation. Also evidence of ancient sausage production exists. The Roman gourmet Apicius speaks of a sausage-making technique involving *œnogaros* (a mixture of the fermented fish sauce *garum* with oil or wine). Preserved meats were furthermore a part of religious traditions: resulting meat for offerings to the gods was salted before being given to priests, after which it could be picked up again by the offerer, or even sold in the butcher's.

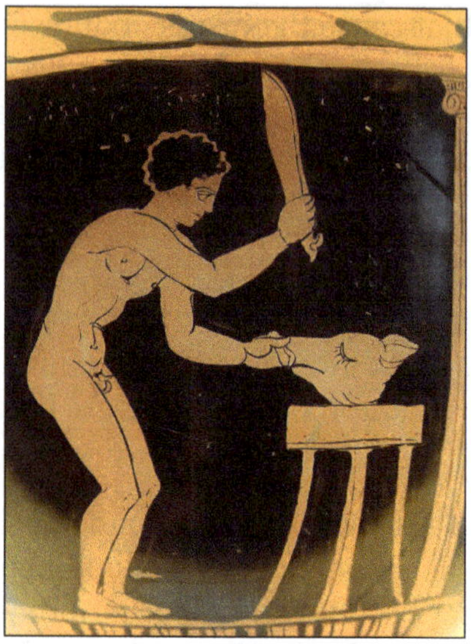

Young man preparing a pig's head after a sacrifice. Vase v. 360-340 BC, National Archaeological Museum of Spain.

A trade in salt meat occurred across ancient Europe. In Polybius's time, the Gauls exported salt pork each year to Rome in large quantities, where it was sold in different cuts: rear cuts, middle cuts, hams, and sausages. This meat, after having been salted with the greatest care, was sometime smoked. These goods had to have been considerably important, since they fed part of the Roman people and the armies. The Belgae were celebrated above all for the care which they gave to the fattening of their pigs. Their herds of sheep and pigs were so many, they could provide skins and salt meat not only for Rome, but also for most of Italy. The Ceretani of Spain drew a large export income from their hams, which were so succulent, they were in no way inferior to those of Cantabria. These *tarichos* of pig would become especially sought, to the point that the ancients considered this meat the most nourishing of all and the easiest to digest.

In Ethiopia, according to Pliny, and in Libya according to Saint Jerome, the Acridophages (literally, the locust-eaters) salted and smoked the crickets which arrived at their settlements in the spring in great swarms and which constituted, it was said, their sole food.

The smoking of meat was a traditional practice in North America, where Plains Indians hung their meat at the top of their tipis to increase the amount of smoke coming into contact with the food.

Middle Ages

In Europe, medieval cuisine made great use of meat and vegetables, and the guild of butchers was amongst the most powerful. During the 12th century, salt beef was consumed by all social classes. Smoked meat was called *carbouclée* in Romance tongues and *bacon* if it was pork.

The Middle Ages made pâté a masterpiece: that which is, in the 21st century, merely spiced minced meat (or fish), baked in a terrine and eaten cold, was at that time composed of a dough envelope stuffed with varied meats and superbly decorated for ceremonial feasts. The first French recipe, written in verse by Gace de La Bigne, mentions in the same pâté three great partridges, six fat quail, and a dozen larks. *Le Ménagier de Paris* mentions pâtés of fish, game, young rabbit, fresh venison, beef, pigeons, mutton, veal, and pork, and even pâtés of lark, turtledove, cow, baby bird, goose, and hen. Bartolomeo Sacchi, called Platine, prefect of the Vatican Library, gives the recipe for a pâté of wild beasts: the flesh, after being boiled with salt and vinegar, was larded and placed inside an envelope of spiced fat, with a *mélange* of pepper, cinnamon and pounded lard; one studded the fat with cloves until it was entirely covered, then placed it inside a pâte.

In the 16th century, the most fashionable pâtés were of woodcock, au bec doré, chapon, beef tongue, cow feet, sheep feet, chicken, veal, and venison. In the same era, Pierre Belon notes that the inhabitants of Crete and Chios lightly salted then oven-dried entire hares, sheep, and roe deer cut into pieces, and that in Turkey, cattle and sheep, cut and minced rouelles, salted then dried, were eaten on voyages with onions and no other preparation.

Early Modern Era

Barrels of salt beef and other products in a reconstruction of an American Civil War stockpile, at Fort Macon State Park, North Carolina.

During the Age of Discovery, salt meat was one of the main foods for sailors on long voyages, for instance in the merchant marine or the navy. In the 18th century, salted Irish beef, transported in barrels, are considered finest.

Scientific research on meat by chemists and pharmacists led to the creation of a new, extremely practical product: meat extract, which could appear in different forms. The need to properly feed soldiers during long campaigns outside the country, such as the Napoleonic Wars, and to nourish a constantly growing population often living in appalling conditions drove scientific research, but a confectioner, Nicolas Appert, in 1795 developed through experimentation a method which would become universal and in one language bears his name: airtight storage, called *appertisation* in French.

With the spread of *appertisation*, the 19th-century world entered the era of the "food industry", which developed new products such as canned salt meat (for example corned beef), but also led to lowered standards of food quality and hygiene – such as those Upton Sinclair described in *The Jungle*. These bad practices led to the creation of the Pure Food and Drug Act in 1906, followed by the national agencies for health security and the establishment of food traceability over the course of the 20th century. It also led to continuing technological innovation.

In France, the summer of 1857 was so hot that most butchers refused to slaughter animals and charcutiers lost considerable amounts of meat, due to inadequate conservation methods. A member of the Academy of Medicine and his son issued a 34-page summary of works printed between 1663 and 1857, which proposed some solutions: not less than 91 texts exist, of which 64 edited for only the years between 1851 and 1857.

Chemical Actions

Salt

Salt (sodium chloride) is the primary ingredient used in meat curing. Removal of water and addition of salt to meat creates a solute-rich environment where osmotic pressure draws water out of microorganisms, slowing down their growth. Doing this requires a concentration of salt of nearly 20%. In addition, salt causes the soluble proteins to come to the surface of the meat that was used to make the sausages. These proteins coagulate when the sausage is heated, helping to hold the sausage together.

Sugar

The sugar added to meat for the purpose of curing it comes in many forms, including honey, corn syrup solids, and maple syrup. However, with the exception of bacon, it does not contribute much to the flavor, but it does alleviate the harsh flavor of the salt. Sugar also contributes to the growth of beneficial bacteria such as *Lactobacillus* by feeding them.

Nitrates and Nitrites

Nitrates and nitrites not only help kill bacteria, but also produce a characteristic flavor and give meat a pink or red color. Nitrite (NO^{-}_{2}) is generally supplied by sodium nitrite or (indirectly) by potassium nitrate. Nitrite salts are most often used in curing. Nitrate is specifically used only in a few curing conditions and products where nitrite (which may be generated from nitrate) must be generated in the product over long periods of time.

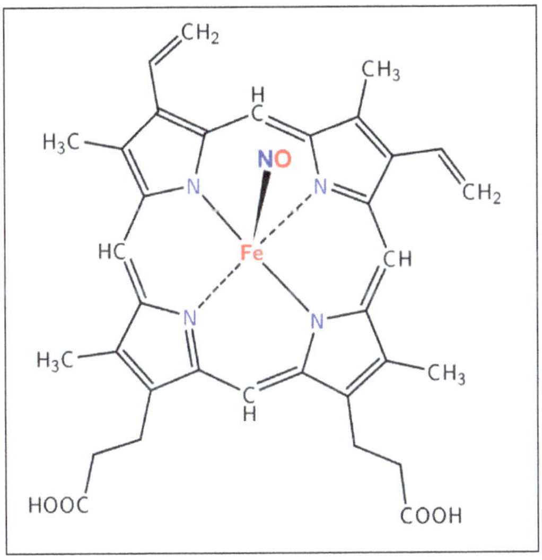

Nitrosyl-heme.

Nitrite further breaks down in the meat into nitric oxide (NO), which then binds to the iron atom in the center of myoglobin's heme group, reducing oxidation and causing a reddish-brown color (nitrosomyoglobin) when raw and the characteristic cooked-ham pink color (nitrosohemochrome or nitrosyl-heme) when cooked. The addition of ascorbate to cured meat reduces formation of nitrosamines, but increases the nitrosylation of iron.

The use of nitrite and nitrate salts for meat in the US has been formally used since 1925. Because of the relatively high toxicity of nitrite (the lethal dose in humans is about 22 mg/kg of body weight), the maximum allowed nitrite concentration in meat products is 200 ppm. Plasma nitrite is reduced in persons with endothelial dysfunction.

The use of nitrites in food preservation is controversial due to the potential for the formation of nitrosamines when nitrites are present in high concentrations and the product is cooked at high temperatures. The effect is seen for red or processed meat, but not for white meat or fish. Nitrates and nitrites may cause cancer and the production of carcinogenic nitrosamines can be potently inhibited by the use of the antioxidants Vitamin C and the alpha-tocopherol form of Vitamin E during curing. Under simulated gastric conditions, nitrosothiols rather than nitrosamines are the main nitroso species being formed. The use of either compound is therefore regulated; for example, in the United States, the concentration of nitrates and nitrites is generally limited to 200 ppm or lower. While the meat industry considers them irreplaceable because of their low cost and efficacy at maintaining color, botulism is an extremely rare disease (less than 1000 cases reported worldwide per year), and almost always associated with home preparations of food storing. Furthermore, while the FDA has set a limit of 200 ppm of nitrates for cured meat, they are not allowed and not recognized as safe in most other foods, even foods that are not cooked at high temperatures, such as cheese.

Nitrites from Celery

Processed meats without "added nitrites" may be misleading as they may be using naturally occurring nitrites from celery instead.

A 2019 report from Consumer Reports found that using celery as a curing agent introduced naturally occurring nitrates and nitrites. The USDA allows the term "uncured" or "no nitrates or nitrites added" on products using these natural sources of nitrites, which provides the consumer a false sense of making a healthier choice. The Consumer Reports investigation also provides the average level of sodium, nitrates and nitrites found per gram of meat in their report.

Consumer Reports and the Center for Science in the Public Interest filed a formal request to the USDA to change the labeling requirements this year.

Smoke

Meat can also be preserved by "smoking". If the smoke is hot enough to slow-cook the meat, this will also keep it tender. One method of smoking calls for a smokehouse with damp wood chips or sawdust. In North America, hardwoods such as hickory, mesquite, and maple are commonly used for smoking, as are the wood from fruit trees such as apple, cherry, and plum, and even corncobs.

Smoking helps seal the outer layer of the food being cured, making it more difficult for bacteria to enter. It can be done in combination with other curing methods such as salting. Common smoking styles include hot smoking, smoke roasting (pit barbecuing) and cold smoking. Smoke roasting and hot smoking cook the meat while cold smoking does not. If the meat is cold smoked, it should be dried quickly to limit bacterial growth during the critical period where the meat is not yet dry. This can be achieved, as with jerky, by slicing the meat thinly.

The smoking of food directly with wood smoke is known to contaminate the food with carcinogenic polycyclic aromatic hydrocarbons.

Effect of Meat Preservation

On Health

Since the 20th century, with respect to the relationship between diet and human disease (e.g. cardiovascular, etc.), scientists have conducted studies on the effects of lipolysis on vacuum-packed or frozen meat. In particular, by analyzing entrecôtes of frozen beef during 270 days at −20 °C (−4 °F), scientists found an important phospholipase that accompanies the loss of some unsaturated fat n-3 and n-6, which are already low in the flesh of ruminants.

In 2015, the International Agency for Research on Cancer of the World Health

Organization classified processed meat, that is, meat that has undergone salting, curing, fermenting, or smoking, as "carcinogenic to humans".

On Trade

Thc improvement of methods of meat preservation, and of the means of transport of preserved products, has notably permitted the separation of areas of production and areas of consumption, which can now be distant without it posing a problem, permitting the exportation of meats.

For example, the appearance in the 1980s of preservation techniques under controlled atmosphere sparked a small revolution in the world's market for sheep meat: the lamb of New Zealand, one of the world's largest exporters of lamb, could henceforth be sold as fresh meat, since it could be preserved from 12 to 16 weeks, which would be a sufficient duration for it to reach Europe by boat. Before, meat from New Zealand was frozen, thus had a much lower value on European shelves. With the arrival of the new "chilled" meats, New Zealand could compete even more strongly with local producers of fresh meat. The use of controlled atmosphere to avoid the depreciation which affects frozen meat is equally useful in other meat markets, such as that for pork, which now also enjoys an international trade.

PICKLING

Pickling is the process of preserving or extending the lifespan of food by either anaerobic fermentation in brine or immersion in vinegar. In East Asia, vinaigrette (vegetable oil and vinegar) is also used as a pickling medium. The pickling procedure typically affects the food's texture, taste and flavor. The resulting food is called a *pickle*, or, to prevent ambiguity, prefaced with *pickled*. Foods that are pickled include vegetables, fruits, meats, fish, dairy and eggs.

A jar of pickled cucumbers (front) and a jar of pickled onions (back).

A distinguishing characteristic is a pH of 4.6 or lower, which is sufficient to kill most bacteria. Pickling can preserve perishable foods for months. Antimicrobial herbs and spices, such as mustard seed, garlic, cinnamon or cloves, are often added. If the food contains sufficient moisture, a pickling brine may be produced simply by adding dry salt. For example, sauerkraut and Korean kimchi are produced by salting the vegetables to draw out excess water. Natural fermentation at room temperature, by lactic acid bacteria, produces the required acidity. Other pickles are made by placing vegetables in vinegar. Like the canning process, pickling (which includes fermentation) does not require that the food be completely sterile before it is sealed. The acidity or salinity of the solution, the temperature of fermentation, and the exclusion of oxygen determine which microorganisms dominate, and determine the flavor of the end product.

When both salt concentration and temperature are low, *Leuconostoc mesenteroides* dominates, producing a mix of acids, alcohol, and aroma compounds. At higher temperatures *Lactobacillus plantarum* dominates, which produces primarily lactic acid. Many pickles start with *Leuconostoc*, and change to *Lactobacillus* with higher acidity.

Process

Bát Tràng porcelain vessel for pickling.

In traditional pickling, fruit or vegetables are submerged in brine (20-40 grams/L of salt (3.2–6.4 oz/imp gal or 2.7–5.3 oz/US gal)), or shredded and salted as in sauerkraut preparation, and held underwater by flat stones layered on top. Alternatively, a lid with an airtrap or a tight lid may be used if the lid is able to release pressure which may result from carbon dioxide buildup. Mold or (white) kahm yeast may form on the surface; kahm yeast is mostly harmless but can impart an off taste and may be removed without affecting the pickling process.

In chemical pickling, the fruits or vegetables to be pickled are placed in a sterilized jar along with brine, vinegar, or both, as well as spices, and are then allowed to mature until the desired taste is obtained.

Pickled cucumbers.

The food can be pre-soaked in brine before transferring to vinegar. This reduces the water content of the food, which would otherwise dilute the vinegar. This method is particularly useful for fruit and vegetables with a high natural water content.

In commercial pickling, a preservative such as sodium benzoate or EDTA may also be added to enhance shelf life. In fermentation pickling, the food itself produces the preservation agent, typically by a process involving *Lactobacillus* bacteria that produce lactic acid as the preservative agent.

Pickled herring.

Alum is used in pickling to promote crisp texture and is approved as a food additive by the United States Food and Drug Administration.

"Refrigerator pickles" are unfermented pickles made by marinating fruit or vegetables in a seasoned vinegar solution. They must be stored under refrigeration or undergo canning to achieve long-term storage.

Pickled mushrooms.

Japanese Tsukemono use a variety of pickling ingredients depending on their type, and are produced by combining these ingredients with the vegetables to be preserved and putting the mixture under pressure.

CANNING

Special-edition steel soup cans commemorating Andy Warhol's paintings.

Canning is a method of preserving food in which the food contents are processed and sealed in an airtight container (jars like Mason jars, and steel and tin cans). Canning provides a shelf life typically ranging from one to five years, although under specific circumstances it can be much longer. A freeze-dried canned product, such as canned dried lentils, could last as long as 30 years in an edible state. In 1974, samples of canned food from the wreck of the *Bertrand*, a steamboat that sank in the Missouri River in 1865, were tested by the National Food Processors Association. Although appearance, smell and vitamin content had deteriorated, there was no trace of microbial growth and the 109-year-old food was determined to be still safe to eat.

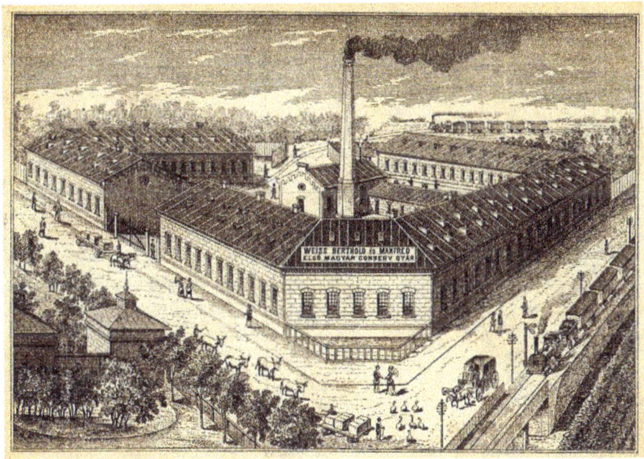

The Berthold-Weiss Factory, one of the first large canned food factories in Csepel-Budapest.

During the first years of the Napoleonic Wars, the French government offered a hefty cash award of 12,000 francs to any inventor who could devise a cheap and effective method of preserving large amounts of food. The larger armies of the period required increased and regular supplies of quality food. Limited food availability was among the factors limiting military campaigns to the summer and autumn months. In 1809, Nicolas Appert, a French confectioner and brewer, observed that food cooked inside a jar did not spoil unless the seals leaked, and developed a method of sealing food in glass jars. Appert was awarded the prize in 1810 by Count Montelivert, a French minister of the interior. The reason for lack of spoilage was unknown at the time, since it would be another 50 years before Louis Pasteur demonstrated the role of microbes in food spoilage.

How canned food was made, Retorts can be seen.

The French Army began experimenting with issuing canned foods to its soldiers, but the slow process of canning foods and the even slower development and transport stages prevented the army from shipping large amounts across the French Empire, and the war ended before the process was perfected.

Following the end of the Napoleonic Wars, the canning process was gradually employed in other European countries and in the US.

Appert canning jar.

Nicolas Appert, developer of the canning process.

Based on Appert's methods of food preservation, the tin can process was allegedly developed by Frenchman Philippe de Girard, who came to London and used British merchant Peter Durand as an agent to patent his own idea in 1810. Durand did not pursue food canning himself, selling his patent in 1811 to Bryan Donkin and John Hall, who were in business as Donkin Hall and Gamble, of Bermondsey. Bryan Donkin developed the process of packaging food in sealed airtight cans, made of tinned wrought iron. Initially, the canning process was slow and labour-intensive, as each large can had to be hand-made, and took up to six hours to cook, making canned food too expensive for ordinary people.

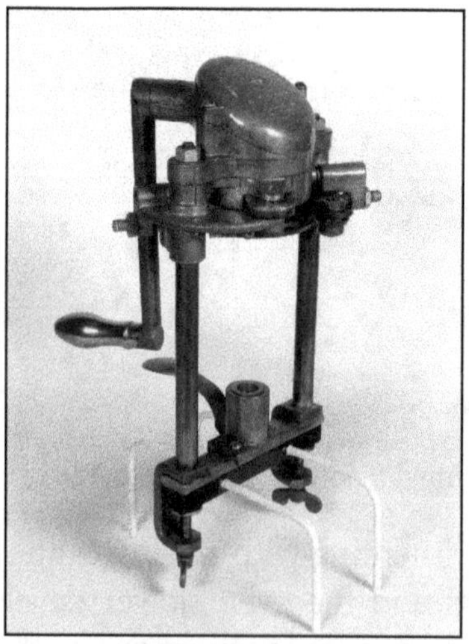

A Dixie Can Sealer for home use. Now in Thinktank, Birmingham Science Museum.

The main market for the food at this stage was the British Army and Royal Navy. By 1817 Donkin recorded that he had sold £3000 worth of canned meat in six months. In 1824 Sir William Edward Parry took canned beef and pea soup with him on his voyage to the Arctic in HMS Fury, during his search for a northwestern passage to India. In 1829, Admiral Sir James Ross also took canned food to the Arctic, as did Sir John Franklin in 1845. Some of his stores were found by the search expedition led by Captain (later Admiral Sir) Leopold McClintock in 1857. One of these cans was opened in 1939, and was edible and nutritious, though it was not analysed for contamination by the lead solder used in its manufacture.

During the mid-19th century, canned food became a status symbol amongst middle-class households in Europe, being something of a frivolous novelty. Early methods of manufacture employed poisonous lead solder for sealing the cans. Studies in the 1980s attributed the lead from the cans as a factor in the disastrous outcome of the 1845 Franklin expedition to chart and navigate the Northwest Passage. Later studies found this to be false.

Increasing mechanization of the canning process, coupled with a huge increase in urban populations across Europe, resulted in a rising demand for canned food. A number of inventions and improvements followed, and by the 1860s smaller machine-made steel cans were possible, and the time to cook food in sealed cans had been reduced from around six hours to thirty minutes.

1914 magazine advertisement for cookware with instructions for home canning.

Canned food also began to spread beyond Europe – Robert Ayars established the first American canning factory in New York City in 1812, using improved tin-plated wrought-iron cans for preserving oysters, meats, fruits and vegetables. Demand for canned food greatly increased during wars. Large-scale wars in the nineteenth century, such as the Crimean War, American Civil War, and Franco-Prussian War introduced increasing numbers of working-class men to canned food, and allowed canning companies to expand their businesses to meet military demands for non-perishable food, allowing companies to manufacture in bulk and sell to wider civilian markets after wars ended. Urban populations in Victorian Britain demanded ever-increasing quantities of cheap, varied, quality food that they could keep at home without having to go shopping daily. In response, companies such as Underwood, Nestlé, Heinz, and others provided quality canned food for sale to working class city-dwellers. In particular, Crosse and Blackwell took over the concern of Donkin Hall and Gamble. The late 19th century saw the range of canned food available to urban populations greatly increase, as canners competed with each other using novel foodstuffs, highly decorated printed labels, and lower prices.

Demand for canned food skyrocketed during World War I, as military commanders sought vast quantities of cheap, high-calorie food to feed their millions of soldiers, which could be transported safely, survive trench conditions, and not spoil in transport. Throughout the war, British soldiers generally subsisted on low-quality canned foodstuffs, such as the British "Bully Beef" (cheap corned beef), pork and beans, canned sausages, and Maconochies Irish Stew, but by 1916, widespread dissatisfaction and increasing complaints about the cheap canned food amongst soldiers resulted in militaries purchasing better-quality food to improve morale, and complete meals-in-a-can began to appear. In 1917, the French Army began issuing canned French cuisine, such as coq au vin, Beef Bourguignon and Vichyssoise while the Italian Army experimented with canned ravioli, spaghetti bolognese, Minestrone and Pasta e fagioli. Shortages of canned food in the British Army in 1917 led to the government issuing cigarettes and amphetamines to soldiers to suppress their appetites. After the war, companies that had supplied military canned food improved the quality of their goods for civilian sale.

Methods

The original fragile and heavy glass containers presented challenges for transportation, and glass jars were largely replaced in commercial canneries with cylindrical tin can or wrought-iron canisters (later shortened to "cans") following the work of Peter Durand. Cans are cheaper and quicker to make, and much less fragile than glass jars. Glass jars have remained popular for some high-value products and in home canning. Can openers were not invented for another thirty years — at first, soldiers had to cut the cans open with bayonets or smash them open with rocks. Today, tin-coated steel is the material most commonly used. Laminate vacuum pouches are also used for canning, such as used in MREs and Capri Sun drinks.

To prevent the food from being spoiled before and during containment, a number of methods are used: pasteurisation, boiling (and other applications of high temperature over a period of time), refrigeration, freezing, drying, vacuum treatment, antimicrobial agents that are natural to the recipe of the foods being preserved, a sufficient dose of ionizing radiation, submersion in a strong saline solution, acid, base, osmotically extreme (for example very sugary) or other microbially-challenging environments.

Other than sterilization, no method is perfectly dependable as a preservative. For example, the microorganism *Clostridium botulinum* (which causes botulism) can be eliminated only at temperatures above the boiling point of water.

From a public safety point of view, foods with low acidity (a pH more than 4.6) need sterilization under high temperature (116–130 °C). To achieve temperatures above the boiling point requires the use of a pressure canner. Foods that must be pressure canned include most vegetables, meat, seafood, poultry, and dairy products. The only foods that may be safely canned in an ordinary boiling water bath are highly acidic ones with a pH below 4.6, such as fruits, pickled vegetables, or other foods to which acidic additives have been added.

Double Seams

Invented in 1888 by Max Ams, modern double seams provide an airtight seal to the tin can. This airtight nature is crucial to keeping micro-organisms out of the can and keeping its contents sealed inside. Thus, double seamed cans are also known as Sanitary Cans. Developed in 1900 in Europe, this sort of can was made of the traditional cylindrical body made with tin plate. The two ends (lids) were attached using what is now called a double seam. A can thus sealed is impervious to contamination by creating two tight continuous folds between the can's cylindrical body and the lids. This eliminated the need for solder and allowed improvements in manufacturing speed, reducing cost.

Double seaming uses rollers to shape the can, lid and the final double seam. To make a sanitary can and lid suitable for double seaming, manufacture begins with a sheet of coated tin plate. To create the can body, rectangles are cut and curled around a die, and welded together creating a cylinder with a side seam.

Rollers are then used to flare out one or both ends of the cylinder to create a quarter circle flange around the circumference. Precision is required to ensure that the welded sides are perfectly aligned, as any misalignment will cause inconsistent flange shape, compromising its integrity.

A circle is then cut from the sheet using a die cutter. The circle is shaped in a stamping press to create a downward countersink to fit snugly into the can body. The result can be compared to an upside down and very flat top hat. The outer edge is then curled down and around about 140 degrees using rollers to create the end curl.

The result is a steel tube with a flanged edge, and a countersunk steel disc with a curled edge. A rubber compound is put inside the curl.

Seaming

Opened can.

The body and end are brought together in a seamer and held in place by the base plate and chuck, respectively. The base plate provides a sure footing for the can body during the seaming operation and the chuck fits snugly into the end (lid). The result is the countersink of the end sits inside the top of the can body just below the flange. The end curl protrudes slightly beyond the flange.

First Operation

Once brought together in the seamer, the seaming head presses a first operation roller against the end curl. The end curl is pressed against the flange curling it in toward the body and under the flange. The flange is also bent downward, and the end and body are now loosely joined together. The first operation roller is then retracted. At this point five thicknesses of steel exist in the seam. From the outside in they are:

- End,
- Flange,
- End Curl,
- Body,
- Countersink.

Second Operation

The seaming head then engages the second operation roller against the partly formed seam. The second operation presses all five steel components together tightly to form the final seal. The five layers in the final seam are then called; a) End, b) Body Hook, c) Cover Hook, d) Body, e) Countersink. All sanitary cans require a filling medium within the seam because otherwise the metal-to-metal contact will not maintain a hermetic seal. In most cases, a rubberized compound is placed inside the end curl radius, forming the critical seal between the end and the body.

Probably the most important innovation since the introduction of double seams is the welded side seam. Prior to the welded side seam, the can body was folded and/or soldered together, leaving a relatively thick side seam. The thick side seam required that the side seam end juncture at the end curl to have more metal to curl around before closing in behind the Body Hook or flange, with a greater opportunity for error.

Seamer Setup and Quality Assurance

Many different parts during the seaming process are critical in ensuring that a can is airtight and vacuum sealed. The dangers of a can that is not hermetically sealed are contamination by foreign objects (bacteria or fungicide sprays), or that the can could leak or spoil.

One important part is the seamer setup. This process is usually performed by an experienced technician. Amongst the parts that need setup are seamer rolls and chucks which have to be set in their exact position (using a feeler gauge or a clearance gauge). The lifter pressure and position, roll and chuck designs, tooling wear, and bearing wear all contribute to a good double seam.

Incorrect setups can be non-intuitive. For example, due to the springback effect, a seam can appear loose, when in reality it was closed too tight and has opened up like a spring. For this reason, experienced operators and good seamer setup are critical to ensure that double seams are properly closed.

Quality control usually involves taking full cans from the line – one per seamer head, at least once or twice per shift, and performing a teardown operation (wrinkle/tightness), mechanical tests (external thickness, seamer length/height and countersink) as well as cutting the seam open with a twin blade saw and measuring with a double seam inspection system. The combination of these measurements will determine the seam's quality.

Use of a statistical process control (SPC) software in conjunction with a manual double-seam monitor, computerized double seam scanner, or even a fully automatic double seam inspection system makes the laborious process of double seam inspection faster and much more accurate. Statistically tracking the performance of each head or seaming station of the can seamer allows for better prediction of can seamer issues, and may be used to plan maintenance when convenient, rather than to simply react after bad or unsafe cans have been produced.

Nutritional Value

Canning is a way of processing food to extend its shelf life. The idea is to make food available and edible long after the processing time. Canned fruits and vegetables are as rich with dietary fiber and vitamins as the same corresponding fresh or frozen foods, and in some cases the canned products are richer than their fresh or frozen counterparts. The heating process during canning appears to make dietary fiber more soluble, and therefore more readily fermented in the colon into gases and physiologically active byproducts. Canned tomatoes have a higher available lycopene content. Consequently, canned meat and vegetables are often among the list of food items that are stocked during emergencies.

PASTEURIZATION

Pasteurization or pasteurisation is a process in which water and certain packaged and non-packaged foods (such as milk and fruit juice) are treated with mild heat, usually to less than 100 °C (212 °F), to eliminate pathogens and extend shelf life. The process is intended to destroy or deactivate organisms and enzymes that contribute to spoilage or risk of disease, including vegetative bacteria, but not bacterial spores. Since pasteurization is not sterilization, and does not kill spores, a second "double" pasteurization will extend the quality by killing spores that have germinated.

The process was named after the French scientist Louis Pasteur, whose research in the 1880s demonstrated that thermal processing would inactivate unwanted microorganisms in wine. Spoilage enzymes are also inactivated during pasteurization. Today, pasteurization is used widely in the dairy industry and other food processing industries to achieve food preservation and food safety.

Most liquid products are heat treated in a continuous system where heat can be applied using a plate heat exchanger and/or the direct or indirect use of hot water and steam. Due to the mild heat, there are minor changes to the nutritional quality and sensory characteristics of the treated foods. Pascalization or high pressure processing (HPP) and pulsed electric field (PEF) are non-thermal processes that are also used to pasteurize foods.

The process of heating wine for preservation purposes has been known in China since AD 1117, and was documented in Japan in the diary *Tamonin-nikki*, written by a series of monks between 1478 and 1618.

Much later, in 1768, Italian priest and scientist Lazzaro Spallanzani's research proved a product could be made "sterile" after thermal processing. Spallanzani boiled meat broth for one hour, sealed the container immediately after boiling, and noticed that the broth did not spoil and was free from microorganisms. In 1795, a Parisian chef and confectioner named Nicolas Appert began experimenting with ways to preserve foodstuffs, succeeding with soups, vegetables, juices, dairy products, jellies, jams, and syrups. He placed the food in glass jars, sealed them with cork and sealing wax and placed them in boiling water. In that same year, the French military offered a cash prize of 12,000 francs for a new method to preserve food. After some 14 or 15 years of experimenting, Appert submitted his invention and won the prize in January 1810. Later that year, Appert published a book. This was the first cookbook of its kind on modern food preservation methods.

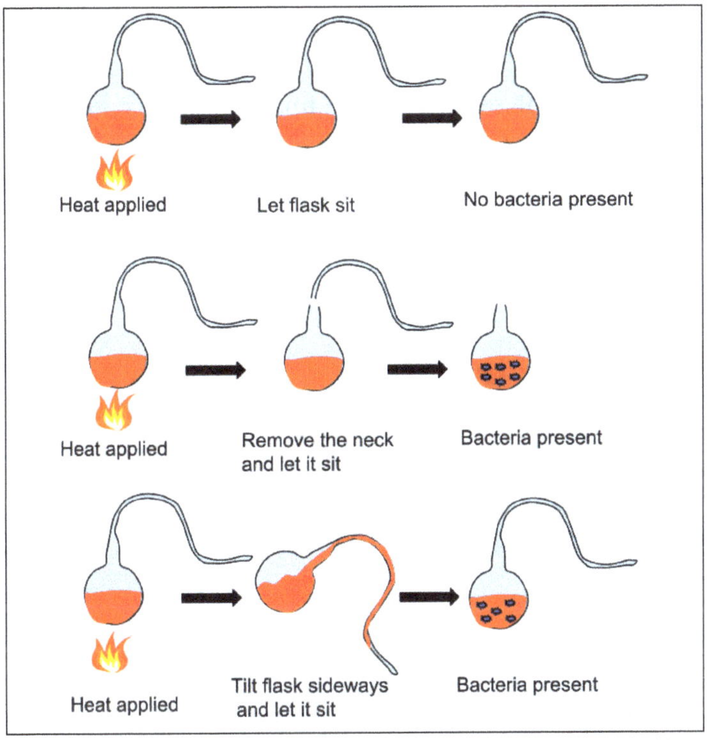

Louis Pasteur's pasteurization experiment illustrates the fact that the spoilage of liquid was caused by particles in the air rather than the air itself. These experiments were important pieces of evidence supporting the idea of the Germ Theory of Disease.

La Maison Appert (English: The House of Appert), in the town of Massy, near Paris, became the first food-bottling factory in the world, preserving a variety of foods in sealed bottles. Appert's method was to fill thick, large-mouthed glass bottles with produce of every description, ranging from beef and fowl to eggs, milk and prepared dishes. He left air space at the top of the bottle, and the cork would then be sealed firmly in the jar by using a vise. The bottle was then wrapped in canvas to protect it while it was dunked into boiling water and then boiled for as much time as Appert deemed appropriate for cooking the contents thoroughly. Appert patented his method, sometimes called *appertisation* in his honor.

Appert's method was so simple and workable that it quickly became widespread. In 1810, British inventor and merchant Peter Durand, also of French origin, patented his own method, but this time in a tin can, so creating the modern-day process of canning foods. In 1812, Englishmen Bryan Donkin and John Hall purchased both patents and began producing preserves. Just a decade later, Appert's method of canning had made its way to America. Tin can production was not common until the beginning of the 20th century, partly because a hammer and chisel were needed to open cans until the invention of a can opener by Robert Yeates in 1855.

A less aggressive method was developed by French chemist Louis Pasteur during an 1864 summer holiday in Arbois. To remedy the frequent acidity of the local aged wines, he found out experimentally that it is sufficient to heat a young wine to only about 50–60 °C (122–140 °F) for a short time to kill the microbes, and that the wine could subsequently be aged without sacrificing the final quality. In honour of Pasteur, this process is known as "pasteurization". Pasteurization was originally used as a way of preventing wine and beer from souring, and it would be many years before milk was pasteurized. In the United States in the 1870s, it was common for milk to contain substances intended to mask spoilage before milk was regulated.

Milk

180 kilograms (400 lb) of milk in a cheese vat.

Milk is an excellent medium for microbial growth, and when it is stored at ambient temperature bacteria and other pathogens soon proliferate. The US Centers for Disease Control (CDC) says improperly handled raw milk is responsible for nearly three times more hospitalizations than any other food-borne disease source, making it one of the world's most dangerous food products. Diseases prevented by pasteurization can include tuberculosis, brucellosis, diphtheria, scarlet fever, and Q-fever; it also kills the harmful bacteria *Salmonella*, *Listeria*, *Yersinia*, *Campylobacter*, *Staphylococcus aureus*, and *Escherichia coli O157:H7*, among others.

Prior to industrialization, dairy cows were kept in urban areas to limit the time between milk production and consumption, hence the risk of disease transmission via raw milk was reduced. As urban densities increased and supply chains lengthened to the distance from country to city, raw milk (often days old) became recognized as a source of disease. For example, between 1912 and 1937, some 65,000 people died of tuberculosis contracted from consuming milk in England and Wales alone. Because tuberculosis has a long incubation period in humans, it was difficult to link unpasteurized milk consumption with the disease. In 1892, chemist Earnest Lederle experimentally inoculated milk from tuberculosis-diseased cows into guinea pigs, which caused them to develop the disease. In 1910, Lederle, then in the role of Commissioner of Health, introduced mandatory pasteurization of milk in New York City.

Developed countries adopted milk pasteurization to prevent such disease and loss of life, and as a result milk is now considered a safer food. A traditional form of pasteurization by scalding and straining of cream to increase the keeping qualities of butter was practiced in Great Britain in the 18th century and was introduced to Boston in the British Colonies by 1773, although it was not widely practiced in the United States for the next 20 years. Pasteurization of milk was suggested by Franz von Soxhlet in 1886. In the early 20th century, Milton Joseph Rosenau established the standards – i.e. low-temperature, slow heating at 60 °C (140 °F) for 20 minutes – for the pasteurization of milk while at the United States Marine Hospital Service. States in the U.S. soon began enacting mandatory dairy pasteurization laws, with the first in 1947, and in 1973 the U.S. federal government required pasteurization of milk used in any interstate commerce.

The shelf life of refrigerated pasteurized milk is greater than that of raw milk. For example, high-temperature, short-time (HTST) pasteurized milk typically has a refrigerated shelf life of two to three weeks, whereas ultra-pasteurized milk can last much longer, sometimes two to three months. When ultra-heat treatment (UHT) is combined with sterile handling and container technology (such as aseptic packaging), it can even be stored non-refrigerated for up to 9 months.

According to the Centers for Disease Control, between 1998 and 2011, 79% of dairy-related disease outbreaks were due to raw milk or cheese products. They report 148 outbreaks and 2,384 illnesses (with 284 requiring hospitalization), as well as two deaths due to raw milk or cheese products during the same time period.

Water

Water is often pasteurized in healthcare facilities as a form of disinfection. The temperature is raised to 70 °C (158 °F) for 30 minutes. Hot water is used to disinfect respiratory and anesthesia equipment. The Centers for Disease Control also recommends heating as a method to destroy intestinal pathogens that may be present in the drinking water in places with inadequate sanitation.

Pasteurization Process

Pasteurization is a mild heat treatment of liquid foods (both packaged and unpackaged) where products are typically heated to below 100 °C. The heat treatment and cooling process are designed to inhibit a phase change of the product. The acidity of the food determines the parameters (time and temperature) of the heat treatment as well as the duration of shelf life. Parameters also take into account nutritional and sensory qualities that are sensitive to heat.

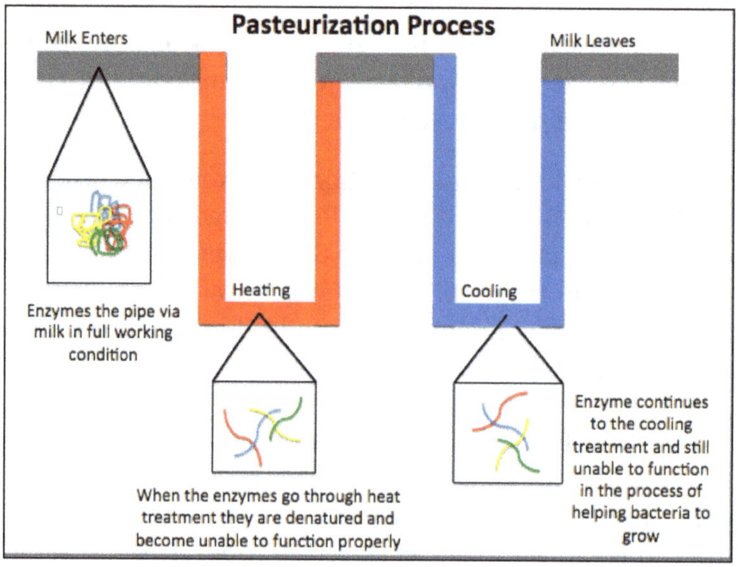

General overview of the pasteurization process. The milk starts at the left and enters the piping with functioning enzymes that, when heat-treated, become denatured and stop the enzymes from functioning. This helps to stop pathogen growth by stopping the functionality of the cell. The cooling process helps stop the milk from undergoing the Maillard reaction and caramelization. The pasteurization process also has the ability to heat the cells to the point that they burst from pressure build-up.

In acidic foods (pH <4.6), such as fruit juice and beer, the heat treatments are designed to inactivate enzymes (pectin methylesterase and polygalacturonase in fruit juices) and destroy spoilage microbes (yeast and lactobacillus). Due to the low pH of acidic foods, pathogens are unable to grow. The shelf-life is thereby extended several weeks. In less acidic foods (pH >4.6), such as milk and liquid eggs, the heat treatments are designed to destroy pathogens and spoilage organisms (yeast and molds). Not all spoilage organisms are destroyed under pasteurization parameters, thus subsequent refrigeration is necessary.

Equipment

Food can be pasteurized in two ways: either before or after being packaged into containers. When food is packaged in glass, hot water is used to lower the risk of thermal shock. Plastics and metals are also used to package foods, and these are generally pasteurized with steam or hot water since the risk of thermal shock is low.

Most liquid foods are pasteurized using continuous systems that have a heating zone, hold tube, and cooling zone, after which the product is filled into the package. Plate heat exchangers are used for low-viscosity products such as animal milks, nut milks and juices. A plate heat exchanger is composed of many thin vertical stainless steel plates which separate the liquid from the heating or cooling medium. Scraped surface heat exchangers contain an inner rotating shaft in the tube, and serve to scrape highly viscous material which might accumulate on the wall of the tube.

Shell or tube heat exchangers are designed for the pasteurization of Non-Newtonian foods such as dairy products, tomato ketchup and baby foods. A tube heat exchanger is made up of concentric stainless steel tubes. Food passes through the inner tube while the heating/cooling medium is circulated through the outer or inner tube.

The benefits of using a heat exchanger to pasteurize non-packaged foods versus pasteurizing foods in containers are:

- Heat exchangers provide uniform treatment, and there is greater flexibility with regards to the products which can be pasteurized on these plates.
- The process is more energy-efficient compared to pasteurizing foods in packaged containers.
- Greater throughput.

After being heated in a heat exchanger, the product flows through a hold tube for a set period of time to achieve the required treatment. If pasteurization temperature or time is not achieved, a flow diversion valve is utilized to divert under-processed product back to the raw product tank. If the product is adequately processed, it is cooled in a heat exchanger, then filled.

High-temperature short-time (HTST) pasteurization, such as that used for milk (71.5 °C (160.7 °F) for 15 seconds) ensures safety of milk and provides a refrigerated shelf life of approximately two weeks. In ultra-high-temperature (UHT) pasteurization, milk is pasteurized at 135 °C (275 °F) for 1–2 seconds, which provides the same level of safety, but along with the packaging, extends shelf life to three months under refrigeration.

Verification

Direct microbiological techniques are the ultimate measurement of pathogen contamination, but these are costly and time-consuming, which means that products have a reduced shelf-life by the time pasteurization is verified.

As a result of the unsuitability of microbiological techniques, milk pasteurization efficacy is typically monitored by checking for the presence of alkaline phosphatase, which is denatured by pasteurization. Destruction of alkaline phosphatase ensures the destruction of common milk pathogens. Therefore, the presence of alkaline phosphatase is an ideal indicator of pasteurization efficacy. For liquid eggs, the effectiveness of the heat treatment is measured by the residual activity of α-amylase.

Efficacy against Pathogenic Bacteria

During the early 20th century, there was no robust knowledge of what time and temperature combinations would inactivate pathogenic bacteria in milk, and so a number of different pasteurization standards were in use. By 1943, both HTST pasteurization conditions of 72 °C (162 °F) for 15 seconds, as well as batch pasteurization conditions of 63 °C (145 °F) for 30 minutes, were confirmed by studies of the complete thermal death (as best as could be measured at that time) for a range of pathogenic bacteria in milk. Complete inactivation of *Coxiella burnetii* (which was thought at the time to cause Q fever by oral ingestion of infected milk) as well as of *Mycobacterium tuberculosis* (which causes tuberculosis) were later demonstrated. For all practical purposes, these conditions were adequate for destroying almost all yeasts, molds, and common spoilage bacteria and also for ensuring adequate destruction of common pathogenic, heat-resistant organisms. However, the microbiological techniques used until the 1960s did not allow for the actual reduction of bacteria to be enumerated. Demonstration of the extent of inactivation of pathogenic bacteria by milk pasteurization came from a study of surviving bacteria in milk that was heat-treated after being deliberately spiked with high levels of the most heat-resistant strains of the most significant milk-borne pathogens.

The mean $\log_{10}$ reductions and temperatures of inactivation of the major milk-borne pathogens during a 15-second treatment are:

- Staphylococcus aureus > 6.7 at 66.5 °C (151.7 °F)
- Yersinia enterocolitica > 6.8 at 62.5 °C (144.5 °F)
- pathogenic Escherichia coli > 6.8 at 65 °C (149 °F)
- Cronobacter sakazakii > 6.7 at 67.5 °C (153.5 °F)
- Listeria monocytogenes > 6.9 at 65.5 °C (149.9 °F)
- Salmonella ser. Typhimurium > 6.9 at 61.5 °C (142.7 °F)

The Codex Alimentarius *Code of Hygienic Practice for Milk* notes that milk pasteurization is designed to achieve at least a 5 $\log_{10}$ reduction of *Coxiella burnetii*. The Code also notes that: "The minimum pasteurization conditions are those having bactericidal effects equivalent to heating every particle of the milk to 72 °C for 15 seconds (continuous flow pasteurization) or 63 °C for 30 minutes (batch pasteurization)" and that "To ensure that each particle is sufficiently heated, the milk flow in heat exchangers should be turbulent, *i.e.* the Reynolds number should be sufficiently high." The point about turbulent flow is important because simplistic laboratory studies of heat inactivation that use test tubes, without flow, will have less bacterial inactivation than larger-scale experiments that seek to replicate conditions of commercial pasteurization.

As a precaution, modern HTST pasteurization processes must be designed with flow-rate restriction as well as divert valves which ensure that the milk is heated evenly and that no part of the milk is subject to a shorter time or a lower temperature. It is common for the temperatures to exceed 72 °C by 1.5 °C or 2 °C.

Effects on Nutritional and Sensory Characteristics of Foods

Because of its mild heat treatment, pasteurization increases the shelf-life by a few days or weeks. However, this mild heat also means there are only minor changes to heat-labile vitamins in the foods.

Milk

According to a systematic review and meta-analysis, it was found that pasteurization appeared to reduce concentrations of vitamins B12 and E, but it also increased concentrations of vitamin A. Apart from meta-analysis, it is not possible to draw conclusions about the effect of pasteurization on vitamins A, B12, and E based merely on consultation of the vast literature available. Milk is not an important source of vitamins B12 or E in the North American diet, so the effects of pasteurization on the adult daily intake of these vitamins is negligible. However, milk is considered an important source of vitamin A, and because pasteurization appears to increase vitamin A concentrations in milk, the effect of milk heat treatment on this vitamin is a not a major public health concern. Results of meta-analyses reveal that pasteurization of milk leads to a significant decrease in vitamin C and folate, but milk is also not an important source of these vitamins. A significant decrease in vitamin B2 concentrations was found after pasteurization. Vitamin B2 is typically found in bovine milk at concentrations of 1.83 mg/liter. Because the recommended daily intake for adults is 1.1 mg/day, milk consumption greatly contributes to the recommended daily intake of this vitamin. With the exception of B2, pasteurization does not appear to be a concern in diminishing the nutritive value of milk because milk is often not a primary source of these studied vitamins in the North American diet.

Sensory Effects

Pasteurization also has a small but measurable effect on the sensory attributes of the foods that are processed. In fruit juices, pasteurization may result in loss of volatile aroma compounds. Fruit juice products undergo a deaeration process prior to pasteurization that may be responsible for this loss. Deaeration also minimizes the loss of nutrients like vitamin C and carotene. To prevent the decrease in quality resulting from the loss in volatile compounds, volatile recovery, though costly, can be utilized to produce higher-quality juice products.

In regards to color, the pasteurization process does not have much effect on pigments such as chlorophylls, anthocyanins and carotenoids in plants and animal tissues. In fruit juices, polyphenol oxidase (PPO) is the main enzyme responsible for causing browning and color changes. However, this enzyme is deactivated in the deaeration step prior to pasteurization with the removal of oxygen.

In milk, the color difference between pasteurized and raw milk is related to the homogenization step that takes place prior to pasteurization. Before pasteurization, milk is homogenized to separate the solids (fat) from the liquid, which results in the pasteurized milk having a whiter appearance compared to raw milk. For vegetable products, color degradation is dependent on the temperature conditions and the duration of heating.

Pasteurization may result in some textural loss as a result of enzymatic and non-enzymatic transformations in the structure of pectin if the processing temperatures are too high as a result. However, with mild heat treatment pasteurization, tissue softening in the vegetables that causes textural loss is not of concern as long as the temperature does not get above 80 °C (176 °F).

Novel Pasteurization Methods

Other thermal and non-thermal processes have been developed to pasteurize foods as a way of reducing the effects on nutritional and sensory characteristics of foods and preventing degradation of heat-labile nutrients. Pascalization or high pressure processing (HPP) and pulsed electric field (PEF) are examples of these non-thermal pasteurization methods that are currently commercially utilized.

Microwave volumetric heating (MVH) is the newest available pasteurization technology. It uses microwaves to heat liquids, suspensions, or semi-solids in a continuous flow. Because MVH delivers energy evenly and deeply into the whole body of a flowing product, it allows for gentler and shorter heating, so that almost all heat-sensitive substances in the milk are preserved.

Low Temperature, Short Time (LTST) is a patented method that implies spraying droplets in a chamber heated below the usual pasteurization temperatures. It takes several thousandth of a second to treat liquid products, so the method is also known as the millisecond technology (MST). It significantly extends the shelf life of products (50+ days) when combined with HTST without damaging the nutrients or flavor. LTST has been commercial since 2019.

FOOD DRYING

Food drying is a method of food preservation in which food is dried (dehydrated or desiccated). Drying inhibits the growth of bacteria, yeasts, and mold through the removal of water. Dehydration has been used widely for this purpose since ancient times; the earliest known practice is 12,000 B.C. by inhabitants of the modern Middle East and Asia regions. Water is traditionally removed through evaporation (air drying, sun drying, smoking or wind drying), although today electric food dehydrators or freeze-drying can be used to speed the drying process and ensure more consistent results.

Flattened fish drying in the sun in Madagascar. Fish are preserved through such traditional methods as drying, smoking and salting.

A whole potato, sliced pieces (right), and dried sliced pieces (left).

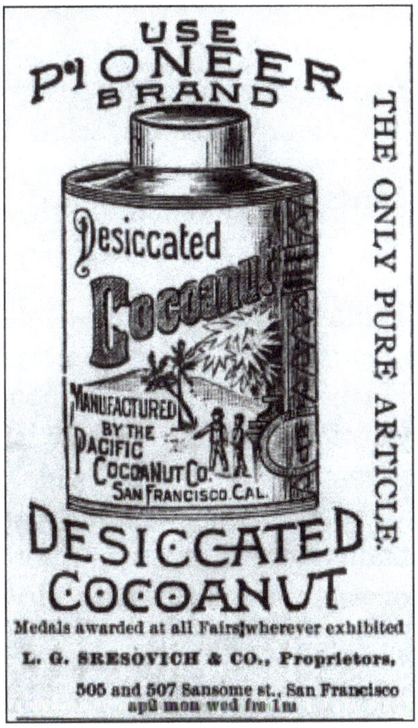

1890 newspaper advertisement showing tin of dried coconut.

Food Types

Many different foods can be prepared by dehydration. Dried fruits have been consumed historically due to their high sugar content and sweet taste, and a longer shelf-life from drying. Fruits may be used differently when dried. The plum becomes a prune, the grape a raisin. Figs and dates may be transformed into different products that can either be eaten as they are, used in recipes, or rehydrated.

A collection of dried mushrooms.

Sun-drying octopus.

Freeze-dried vegetables are often found in food for backpackers, hunters, and the military. Garlic and onion are often dried. Edible mushrooms, as well as other fungi, are also sometimes dried for preservation purposes or to be used as seasonings.

Preparation

Home drying of vegetables, fruit and meat can be carried out with electrical dehydrators (household appliance) or by sun-drying or by wind. Preservatives such as potassium metabisulfite, BHA, or BHT may be used, but are not required. However, dried products without these preservatives may require refrigeration or freezing to ensure safe storage for a long time.

Industrial food dehydration is often accomplished by freeze-drying. In this case food is flash frozen and put into a reduced-pressure system which causes the water to sublimate directly from the solid to the gaseous phase. Although freeze-drying is more expensive than traditional dehydration techniques, it also mitigates the change in flavor, texture, and nutritional value.In addition, another widely used industrial method of drying of food is convective hot air drying. Industrial hot air dryers are simple and easy to design, construct and maintain. More so, it is very affordable and has been reported to retain most of the nutritional properties of food if dried using appropriate drying conditions.

Another form of food dehydration is irradiation. Irradiation uses x-rays, ultraviolet light, and ionizing radiations to penetrate food to the point of sterilization. Astronauts and people who are highly at risk for microbial infections benefit from this method of food drying.

Hurdle technology is the combination of multiple food preservation methods. Hurdle technology uses low doses of multiple food preservation techniques in order to ensure food is not only safe but is desirable visually and texturally.

Packaging

Packaging ensures effective food preservation. Some methods of packaging that are beneficial to dehydrated food are vacuum sealed, inert gases, or gases that help regulate respiration, biological organisms, and growth of microorganisms.

Other Methods

This electric food dehydrator has a hot air blower that blows air through trays with foods on them. Pictured are mango and papaya slices being dried.

There are many different methods for drying, each with their own advantages for particular applications. These include:

- Convection drying.
- Bed dryers.
- Drum drying.
- Freeze Drying.
- Microwave-vacuum drying.
- Shelf dryers.
- Spray drying.
- Infrared radiation drying.
- Combined thermal hybrid drying.
- Sunlight.
- Commercial food dehydrators.
- Household oven.

BIOPRESERVATION

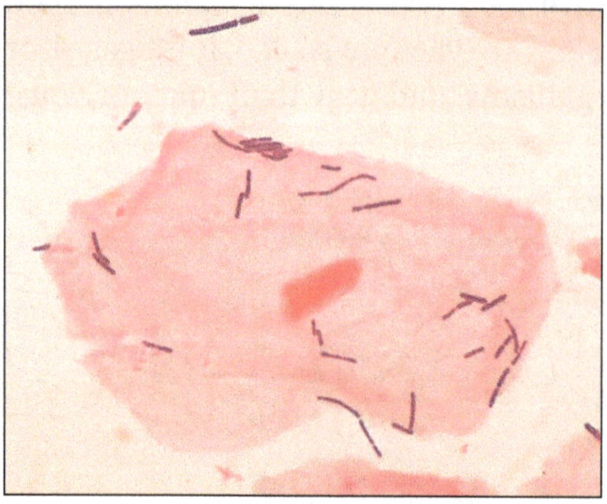

The small rods shown here are lactic acid bacteria which convert lactose and other sugars to lactic acid. The products of their metabolism can have benign preservative effects.

Biopreservation is the use of natural or controlled microbiota or antimicrobials as a way of preserving food and extending its shelf life. The biopreservation of food, especially utilizing lactic acid bacteria (LAB) that are inhibitory to food spoilage microbes, has been practiced since early ages, at first unconsciously but eventually with an increasingly robust scientific foundation. Beneficial bacteria or the fermentation products produced by these bacteria are used in biopreservation to control spoilage and render pathogens inactive in food. There are a various modes of action through which microorganisms can interfere with the growth of others such as organic acid production, resulting in a reduction of pH and the antimicrobial activity of the un-dissociated acid molecules, a wide variety of small inhibitory molecules including hydrogen peroxide, etc. It is a benign ecological approach which is gaining increasing attention.

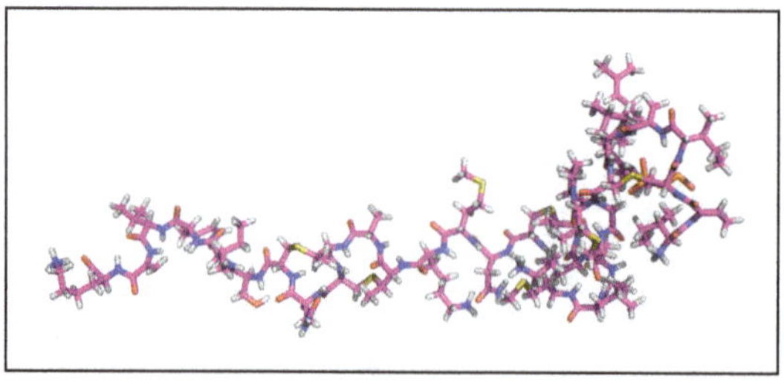

3D stick model of nisin, a particularly effective preservative produced by some lactic acid bacteria.

Biopreservative Agents and Modes of Action

Lactic Acid Bacteria

Of special interest are lactic acid bacteria (LAB). Lactic acid bacteria have antagonistic properties which make them particularly useful as biopreservatives. When LABs compete for nutrients, their metabolites often include active antimicrobials such as lactic and acetic acid, hydrogen peroxide, and peptide bacteriocins. Some LABs produce the antimicrobial nisin which is a particularly effective preservative. These days LAB bacteriocins are used as an integral part of hurdle technology. Using them in combination with other preservative techniques can effectively control spoilage bacteria and other pathogens, and can inhibit the activities of a wide spectrum of organisms, including inherently resistant Gram-negative bacteria." Lactic acid bacteria and propionibacteria have been extensively studies for their efficacy against spoilage causing yeasts and molds in food spoilage.

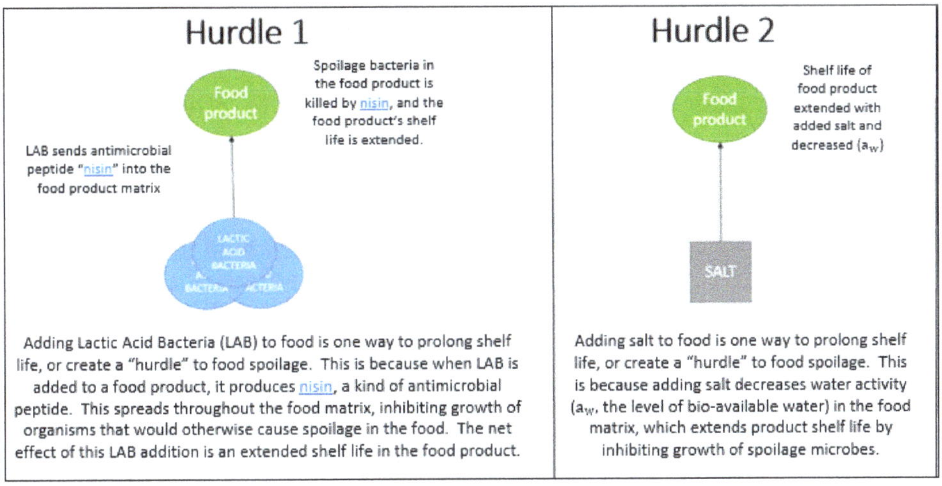

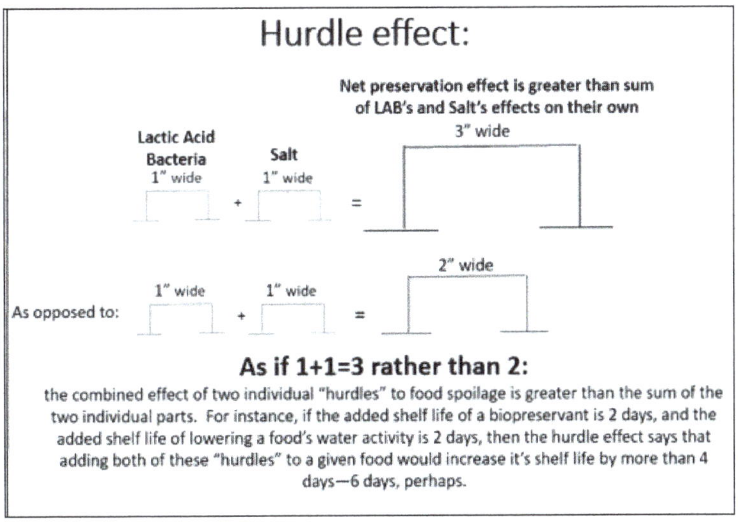

This figure illustrates the pathway of food preservation followed by lactic acid bacteria involving nisin, as well as the pathway of food preservation followed by salt. Additionally, the hurdle effect of food preservation, such as by adding lactic acid bacteria and salt to a food product, is illustrated and described.

Yeast

In addition to lactic acid bacteria, yeasts also have been reported to have a biopreservation effect due to their antagonistic activities relying on the competition for nutrients, production and tolerance of high concentrations of ethanol, as well as the synthesis of a large class of antimicrobial compounds exhibiting large spectrum of activity against food spoilage microorganisms, but also against plant, animal and human pathogen.

A bacterium/yeast that is a suitable candidate for use as a biopreservative does not necessarily have to ferment the food. However, if conditions are suitable for microbial growth, then a biopreservative bacterium will compete well for nutrients with the spoilage and pathogenic bacteria in the food. As a product of its metabolism, it should also produce acids and other antimicrobial agents, particularly bacteriocins. Biopreservative bacteria, such as lactic acid bacteria, must be harmless to humans.

Bacteriophages

Bacteriophages, or simply phages, are viruses which infect bacteria. The majority of all bacteriophages known exhibit a double-stranded DNA genome inside the virion capsid and belong to the order of tailed phages, Caudovirales. The tailed phages can be further separated into three families: Podoviridae, which are characterized by very short tails; Myoviridae, which exhibit longer, straight and contractile tails; and Siphoviridae, which can be identified due to their long and flexible tails. Another well studied group of phages with many applications, although minor in terms of species diversity, is represented by filamentous phages which exhibit a single stranded DNA genome decorated by a helical protein layer surrounding the DNA molecule. Bacteriophages are ubiquitously distributed in nature and can also be isolated from human or animal associated microflora. They outnumber their bacterial host species by a factor of ten representing the most abundant self-replicating entities on earth with an estimated 1031 phages in total. The idea of using phages against unwanted bacteria developed shortly after their discovery. With the improvements in organic chemistry during the 1950s, exploration and development of broad spectrum antibiotics displaced interest in bacteriophage research. Several laboratories have been testing suitability of bacteriophage isolates to control certain bacterial pathogens. Significant advancements in this research have been made at the Bacteriophage Institute in Tbilisi, Georgia, where phage therapy is routinely applied in medicine research field. Today treatment of antibiotic resistant bacteria is a challenging task. Recently, research on bacteriophages has gained additional momentum in light of the identification of antibiotic-resistant pathogens of infectious diseases, wherein the application of antibiotics is not effectively working, therefore research on the application of bacteriophages is being reviewed intensely. Bacteriophages have recently received a generally recognized as safe status based on their lack of toxicity and other detrimental effects to human health for application in meat products in USA.

Phage preparations specific for *L. monocytogenes, E. coli* O157:H7, and *S. enterica* serotypes have been commercialized and approved for application in foods or as part of surface decontamination protocols.

Meat Biopreservation

In meat processing, biopreservation has been extensively studied in fermented meat products and ready to eat meat products. The use of native or artificially-introduced microbial population to improve animal health and productivity, and/or to reduce pathogenic organisms, has been termed a probiotic or competitive enhancement approach. Competitive enhancement strategies that have been developed include competitive exclusion, addition of a microbial supplement (probiotic) that improves gastrointestinal health, and adding a limiting, non-host digestible nutrient (prebiotic) that provides an existing (or introduced) commensal microbial population a competitive advantage in the gastrointestinal tract. Each of these approaches utilizes the activities of the native microbial ecosystem against pathogens by capitalizing on the natural microbial competition. Generally speaking, competitive enhancement strategies offer a natural 'green' method to reduce pathogens in the gut of food animals.

Seafood Biopreservation

Fishery products are a source of wide variety of valuable nutrients such as proteins, vitamins, minerals, omega-3 fatty acids, taurine, etc. Fishery products, however, are also associated with human intoxication and infection. Approximately 10 to 20% of food-borne illnesses are attributed to fish consumption. Changing consumer demand has driven the appeal of traditional processes applied to seafood (e.g. salting, smoking and canning) lower compared to mild technologies involving lower salt content, lower cooking temperature and vacuum packing (VP)/modified atmosphere packing (MAP). These products, designed as lightly preserved fish products (LPFP), are usually produced from fresh seafood and further processing increases risk of cross contamination. These milder treatments are usually not sufficient to destroy microorganisms, and in some cases psychrotolerant pathogenic and spoilage bacteria can develop during the extended shelf-life of LPFP. Many of these products are also eaten raw, so minimizing the presence and preventing growth of microorganisms is essential for the food quality and safety. The microbial safety and stability of food are based on an application of preservative factors called hurdles. The delicate texture and flavor of seafood are very sensitive to the decontamination technologies such as cooking, and more recent mild technologies such as pulsed light, high pressure, ozone, and ultrasound. Chemical preservatives, which are not processes but ingredients, are out of favor with consumers due to natural preservatives demand. An alternative solution that is gaining more and more attention is biopreservation technology. In fish processing, biopreservation is achieved by adding antimicrobials or by increasing the acidity of the fish muscle. Most bacteria stop multiplying when the pH is less than 4.5. Traditionally, acidity has been increased by fermentation, marination or by directly adding acetic, citric or lactic acid to food products. Other preservatives include nitrites, sulphites, sorbates, benzoates and essential oils. The main reason for less documented studies for application of protective microorganisms, bacteriophages or bacteriocins on seafood products for biopreservation compared to dairy or meat products is probably that the early stages of biopreservation have occurred mainly in fermented foodstuffs that are not so developed among seafood products. The selection of potential protective bacteria in seafood products is challenging due to the fact that they need adaptation to the seafood matrix (poor in sugar and their metabolic activities should not change the initial characteristics of the product, i.e. by acidification, and not induce spoilage that could lead to a sensory rejection. Among the microbiota identified in fresh or processed seafood, LAB remains the category that offers the highest potential for direct application as a bioprotective culture or for bacteriocin production.

Commercial Applications and Products

There has been successful implementation of various phage preparation around the globe. Various applications/delivery methods in food have been developed. Bacteriophages and their endolysins can be incorporated into food systems in several ways such as spraying, dipping or immobilization, singly or in combination with other hurdles. The phage preparation LMP-102 has been subsequently commercialized as "List-Shield" Intralyx, Inc. It has been shown to be effective against 170 different strains of "*L. monocytogenes*", reducing significantly (10 to 1000-fold) the Listeria contamination when sprayed onto ready-to-eat foods, without changing the food general composition, taste, odor or color. The Intralytix company has also commercialized phage-based antimicrobial preparations like SalmoFresh and SalmoLyse for controlling *S. enterica*. SalmoFresh is prepared with a cocktail of naturally occurring lytic bacteriophages that selectively and specifically kill Salmonella, including strains belonging to the most common/highly pathogenic serotypes Typhimurium, Enteritidis, Heidelberg, Newport, Hadar, Kentucky and Thompson. According to the manufacturer, SalmoFresh is specifically designed for treating foods that are at high risk for "*Salmonella*" contamination. In particular, red meat and poultry can be treated prior to grinding for significant reductions in Salmonella contamination. SalmoLyse is a reformulated phage cocktail derived from SalmoFresh in which two of the six phages in the original cocktail have been replaced. Additional bacteriophage preparations have been formulated and referenced to be used to reduce the microbial load of animals prior to slaughter and are commercially available from Omnilytics such as the BacWash product line against Salmonella Omnilytics. Another commercial application has been developed, Listex P100 by Micreos in The Netherlands and was granted generally recognized as safe (Gras) status by the Fda and Usda for use in all food products.

Another significant commercial bacteriophage application is Elicosali, a wide range of anti-*Salmonella* and "*E. coli*" phage cocktail, for treatment of agricultural products developed by Eliava Institute at Tbilisi, Republic of Georgia Eliava Institute.

Safety

Biopreservation judiciously exploits the antimicrobial potential of naturally occurring microorganisms in food and their metabolites with a long history of safe use. Bacteriocins, bacteriophages and bacteriophage-encoded enzymes fall in this theory. The long and traditional role of Lactic acid bacteria on food and feed fermentations is the main factor related to the use of bacteriocins in biopreservation. LAB and their bacteriocins have been consumed unintentionally for ages, laying down a long history of safe use. Their antimicrobial spectrum of inhibition, bactericidal mode of action, relative tolerance to processing conditions (pH, NaCl, heat treatments) and the lack of toxicity towards eukaryotic cells enforces their role as biopreservatives in food. The evaluation of any new antimicrobial actives is done in meat by USDA which relies on the GRAS assessment by FDA among other suitability data.

OTHER METHODS OF FOOD PRESERVATION

Pasteurization and Appertization

Foods are subject to thermal processes in a number of different contexts. Often, their main objective is not destruction of micro-organisms in the product, although this is an inevitable and frequently useful side effect.

Heat Process applied to Foods

Heat process	Temperature	Objective
Cooking Baking Boiling Frying Grilling	<100 °Cq	Improvement of digestibility, e.g. Starch gelatinization, Collagen breakdown during Cooking of meat Improvement of flavour Destruction of pathogenic micro organisms
Blanching	<100 °C	Expulsion of oxygen from tissues inactivation of enzymes
Drying/Concentration	<100 °C	Removal of water to enhance keeping quality
Pasteurization	60-80 °C	Elimination of key pathogens and spoilage organisms
Appertization	>100 °C	Elimination of micro-organisms to achieve commercial sterlity

Credit for discovering the value of heat as a preservative agent goes to the French chef, distiller and confectioner, Nicolas Appert. In 1795 the French Directory offered a prize of 12 000 francs to anyone who could develop a new method of preserving food.

Appert won this prize in 1810 after he had experimented for a number of years to develop a technique based on packing foods in glass bottles, sealing them, and then heating them in boiling water. He described his technique in detail in 1811 in a book called the 'Art of Conserving all kinds of Animal and Vegetable Matter for several Years'.

A similar technique was used by the Englishman Saddington in 1807 to preserve fruits and for which he too received a prize, this time of five guineas from the Royal Society of Arts. British patents describing the use of iron or metal containers were issued to Durand and de Heine in 1810 and the firm of Donkin and Hall established a factory for the production of canned foods in Bermondsey, London around 1812.

Appert held the view that the cause of food spoilage was contact with air and that the success of his technique was due to the exclusion of air from the product. This view persisted with sometimes disastrous consequences for another 50 years until Pasteur's work established the relationship between microbial activity and putrefaction. Today, the two types of heat process employed to destroy microorganisms in food, pasteurization and appertization, bear the names of these eminent figures.

Pasteurization, the term given to heat processes typically in the range 60-80 °C and applied for up to a few minutes, is used for two purposes. First is the elimination of a specific pathogen or pathogens associated with a product.

This type of pasteurization is often a legal requirement introduced as a public health measure when a product has been frequently implicated as a vehicle of illness. Notable examples are milk, bulk liquid egg and ice cream mix, all of which have a much improved safety record as a result of pasteurization.

The second reason for pasteurizing a product is to eliminate a large proportion of potential spoilage organisms, thus extending its shelf-life. This is normally the objective when acidic products such as beers, fruit juices, pickles, and sauces are pasteurized.

Where pasteurization is introduced to improve safety, its effect can be doubly beneficial. The process cannot discriminate between the target pathogen(s) and other organisms with similar heat sensitivity so a pasteurization which destroys say Salmonella will also improve shelf-life.

The converse does not normally apply since products pasteurized to improve keeping quality are often intrinsically safe due to other factors such as low pH. This may be less true in the future, however, following the trend toward less acidic and minimally processed foods.

On its own, the contribution of pasteurization to extension of shelf-life can be quite small, particularly if the pasteurized food lacks other contributing preservative factors such as low pH or Thermoduric organisms such as spore formers and some Gram-positive vegetative species in the genera Enterococcus, Micro-bacterium and Arthrobacter can survive pasteurization temperatures. They can also grow and spoil a product quite rapidly at ambient temperatures, so refrigerated storage is often an additional requirement for an acceptable shelf-life.

Appertization refers to processes where the only organisms that survive processing are non-pathogenic and incapable of developing within the product under normal conditions of storage. As a result, appertized products have a long shelf-life even when stored at ambient temperatures.

The term was coined as an alternative to the still widely used description commercially sterile which was objected to on the grounds that sterility is not a relative concept; a material is either sterile or it is not. An appertized or commercially sterile food is not necessarily sterile — completely free from viable organisms.

It is however free from organisms capable of growing in the product under normal storage conditions. Thus for a canned food in temperate climates, it is not a matter of concern if viable spores of a thermophile are present as the organism will not grow at the prevailing ambient temperature.

Aseptic Packaging

Up until now in our consideration of appertized foods we have discussed only retorted products; those which are hermetically sealed into containers, usually cans, and then subjected to an appertizing heat process in-pack. While this has been hugely successful as a long term method of food preservation, it does require extended heating periods in which a food's functional and chemical properties can be adversely affected.

In UHT processing the food is heat processed before it is packed and then sealed into sterilized containers in a sterile environment. This approach allows more rapid heating of the product, the use of higher temperatures than those employed in canning, typically 130-140 °C, and processing times of seconds rather than minutes.

The advantage of using higher temperatures is that the £ value for chemical reactions such as vitamin loss, browning reactions and enzyme inactivation is typically 25- 40 °C compared with 10 °C for spore inactivation. This means that they are less temperature sensitive so that higher temperatures will increase the microbial death rate more than they increase the loss of food quality associated with thermal reactions.

F_0 values for UHT processes can be estimated from the holding temperature (T) and the residence time of the fastest moving stream of product, t.

$$F_0 = 10^{(T - 121/10)} \cdot t \quad (412)$$

Initially UHT processing and aseptic packaging were confined to liquid products such as milk, fruit juices and some soups which would heat up very quickly due to convective heat transfer. If a food contained solid particles larger than about 5 mm diameter it was unsuited to the rapid processing times due to the slower conductive heating of the particulate phase. Scraped surface heat exchangers have been used to process products containing particles up to 25 mm in diameter but at the cost of over-processing the liquid phase.

To avoid this, one system processes the liquid and solid phase separately. A promising alternative is the use of ohmic heating in which a food stream is passed down a tube which contains a series of electrodes. An alternating voltage is applied across the electrodes and the food's resistance causes it to heat up rapidly. Most of the energy supplied is transformed into heat and the rate at which different components heat up is determined by their conductivities rather than heat transfer.

A common packing system used in conjunction with UHT processing is a form/fill/seal operation in which the container is formed in the packaging machine from a reel of plastic or laminate material, although some systems use preformed containers. Packaging is generally refractory to microbial growth and the level of contamination on it is usually very low. Nevertheless to obtain commercial sterility it is given a bactericidal treatment, usually with hydrogen peroxide, sometimes coupled with UV irradiation.

Irradiation

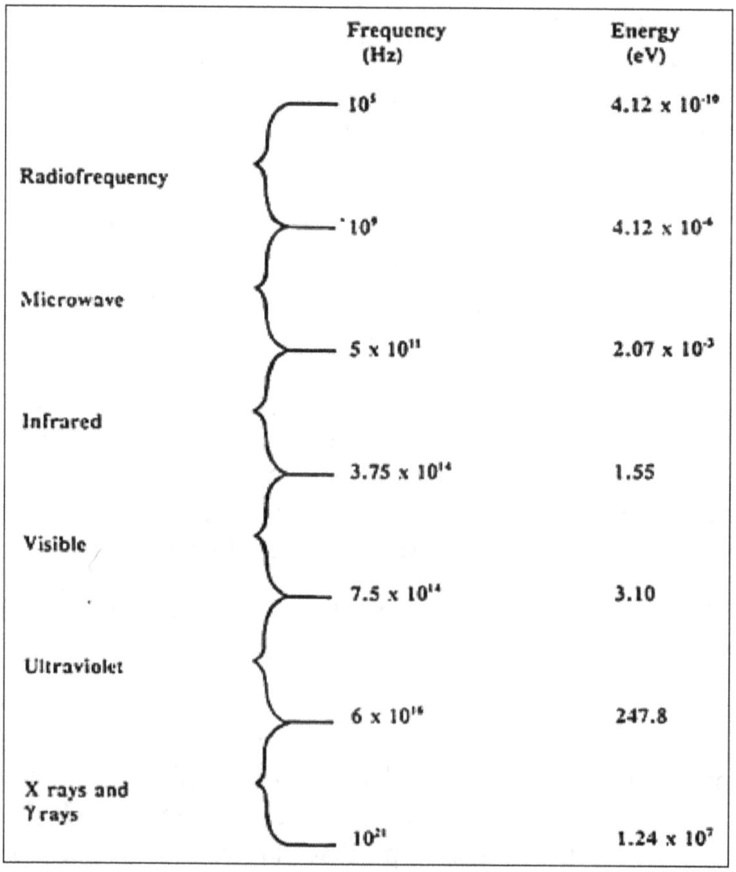

The electromagnetic spectrum.

Electromagnetic (e.m.) radiation is a way in which energy can be propagated through space. It is characterized in terms of its wavelength A, or its frequency v, and the product of these two properties gives the speed, c, at which it travels (3×10^8 m sec^{-1} in a vacuum).

$$v = c \quad (1.13)$$

The range of frequencies (or wavelengths) that e.m. radiation can have is known as the electromagnetic spectrum and is grouped into a number of regions, visible light being only one small region.

The energy carried by e.m. radiation is not continuous but is transmitted in discrete packets or quanta; the energy, E, contained in each quantum being given by the expression:

$$E = hv \quad (4.14)$$

where h is a constant (6.6×10^{-27} ergs sec^{-1}) known as Planck's constant. Thus, the higher the frequency of the radiation the higher its quantum energy.

As far as food microbiology is concerned, only three areas of the e.m. spectrum concern us; microwaves, the UV region and gamma rays.

High-Pressure Processing – Pascalization

Hite showed that high hydrostatic pressures, around 650 MPa (6500 atm), reduced the microbial load in foods such as milk, meats and fruits.

He found that 680 MPa applied for 10 min at room temperature reduced the viable count of milk from 10^7cfu ml^{-1} to 10^1-10″ cfu ml^{-1} and that peaches and pears subjected to 410 MPa for 30 min remained in good condition after 5 years storage. He also noted that the microbicidal activity of high pressure is enhanced by low pH or temperatures above and below ambient.

Since then, microbiologists have continued to study the effect of pressure on micro-organisms, although this work has centred on organisms such as those growing in the sea at great depths and pressures. Interest in the application of high pressures in food processing, sometimes called pascalization, lapsed until the 1980s when progress in industrial ceramic processing led to the development of pressure equipment capable of processing food on a commercial scale and a resurgence of interest, particularly in Japan.

High hydrostatic pressure acts primarily on non-covalent linkages, such as ionic bonds, hydrogen bonds and hydrophobic interactions, and it promotes leactions in which there is an overall decrease in volume. It can have profound effects on proteins, where such interactions are critical to structure and function, although the effect is variable and depends on individual protein structure.

Some proteins such as those of egg, meat and soya form gels and this has been employed to good effect in Japan where high pressure has been used to induce the gelation of fish proteins in the product surimi. Other proteins are relatively unaffected and this can cau6e problems when they have enzymic activity which limits product shelf-life.

Pectin esterase in orange juice, for instance, must be inactivated to stabilize the desired product cloudiness but is very stable to pressures up to 1000 MPa. Non-protein macromolecules can also be affected by high pressures so that pascalized starch products often taste sweeter due to conformational changes in the starch which allow salivary amylase greater access.

Adverse effects on protein structure and activity obviously contribute to the antimicrobial effect of high pressures, although the cell membrane also appears to be an important target. Membrane lipid bilayers have been shown to compress under pressure and this alters their permeability. As a general rule vegetative bacteria and fungi can be reduced by at least one log cycle by 400 MPa applied for 5 min.

Bacterial endospores are more resistant to hydrostatic pressure, tolerating pressures as high as 1200 MPa. Their susceptibility can be increased considerably by modest increases in temperature, when quite low pressures (100 MPa) can produce spore germination, a process in which the spores lose their resistance to heat and to elevated pressure.

Hydrostatic processing has a number of appealing features for the food technologist. It acts instantly and uniformly throughout a food so that the processing time is not related to container size and there are none of the penetration problems associated with heat processing. With the exceptions noted above, adverse effects on the product are slight; nutritional quality, flavour, appearance and texture resemble the fresh material very closely. To the consumer it is a 'natural' process with none of the negative associations of processes such as irradiation or chemical preservatives.

At present, commercial application of high-pressure technology has been limited to acidic products. The yeasts and moulds normally responsible for spoilage in these products are pressure sensitive and the bacterial spores that survive processing are unable to grow at the low pH. In 1990, the Meidi-Ya company in Japan launched a range of jams treated at 400-500 MPa in pack.

These have a chill shelf-life of 60 days and have sensory characteristics quite different from conventional heat- processed jams since more fresh fruit flavour and texture are retained. Other products introduced include salad dressings, fruit sauces, and fruit flavoured yoghurts.

In the future, the range of products may be increased by coupling moderate pressure with a heat treatment equivalent to pasteurization. In one trial, shelf stable, low acid foods were produced by combining a pressure of just 0.14 MPa with heating at temperatures of 82-103 °C. Other developments such as equipment capable of semi- or fully-continuous operation will also considerably improve commercial feasibility, so that we may see and hear a lot more about pascalization.

Low-Temperature Storage – Chilling and Freezing

The rates of most chemical reactions are temperature dependent; as the temperature is lowered so the rate decreases. Since food spoilage is usually a result of chemical reactions mediated by microbial and endogenous enzymes, the useful life of many foods can be increased by storage at low temperatures.

Though this has been known since antiquity, one of the earliest recorded experiments was conducted by the English natural philosopher Francis Bacon who in 1626 stopped his coach in High-gate in order to fill a chicken carcass with snow to confirm that it delayed putrefaction.

This experiment is less notable for its results, which had no immediate practical consequences, than for its regrettable outcome. As a result of his exertions in the snow, it is claimed Bacon caught a cold which led to his death shortly after.

Using low temperatures to preserve food was only practicable where ice was naturally available. As early as the 11th Century BC the Chinese had developed ice houses as a means of storing ice through the summer months, and these became a common feature of large houses in Europe and North America in the 17th and 18th Centuries. By the 19th Century, the cutting and transporting of natural ice had become a substantial industry in areas blessed with a freezing climate.

Mechanical methods of refrigeration and ice making were first patented in the 1830s. These were based on the cooling produced by the vaporization of refrigerant liquids, originally ether but later liquid ammonia. Much early development work was done in Australia where there was considerable impetus to find a way of transporting the abundant cheap meat available locally to European population centres. At the 1872 Melbourne Exhibition, Joseph Harrison exhibited an 'ice house' which kept beef and mutton carcasses in good condition long enough for some of it to be eaten at a public luncheon the following year.

This banquet was to send off a steamship to London carrying 20 tons of frozen mutton and beef packed in tanks cooled by ice and salt. Unfortunately it was an inauspicious start, during passage through the tropics the ice melted and most of the meat had been thrown overboard before the ship reached London. Chilled rather than frozen meat had however already been successfully shipped the shorter distance from North America to Europe and by the end of the Century techniques had been refined to the extent that shipping chilled and frozen meat from North and South America and Australia to Europe was a large and profitable enterprise.

Since then, use of chilling and freezing has extended to a much wider range of perishable foods and to such an extent that refrigeration is now arguably the technology of paramount importance to the food industry.

Chemical Preservatives

The addition of chemicals to food is not a recent innovation but has been practiced throughout recorded history. Doubtless too, there has also always been a certain level of misuse but this must have gone largely undetected until modern analytical techniques became available.

When chemical analysis and microscopy were first applied to foods in the early 19th Century, they revealed the appalling extent of food adulteration then current.'A Treatise on Adulterations of Food, and Culinary Poisons' by Frederick Accum in 1820 marks a watershed.

Accum exposed a horrifying range of abuses such as the sale of sulfuric acid as vinegar, the use of copper salts to colour pickles, the use of alum to whiten bread, addition of acorns to coffee, blackthorn leaves to tea, cyanide to give wines a nutty flavour and red lead to colour Gloucester cheese.

These and subsequent investigations, is directly to the introduction of the first British Food and Drugs Act in 1860. Despite the protection of a much stricter regulatory framework, occasional triumphs of human cupidity are still recorded today.

Recent examples include the use of ethylene glycol in some Austrian wines, the intrepid entrepreneur who sold grated umbrella handles as Parmesan cheese and the grim case of the Spanish toxic cooking oil scandal which killed or maimed hundreds.

Although some would regard all chemical additions to food as synonymous with adulteration, many are recognized as useful and are allowed. Additives may be used to aid processing, to modify a food's texture, flavour, nutritional quality or colour but, here, we are concerned with those which primarily effect keeping quality: preservatives.

Preservatives are defined as 'substances capable of inhibiting, retarding or arresting the growth of micro-organisms or of any deterioration resulting from their presence or of masking the evidence of any such deterioration'.

They do not therefore include substances which act by inhibiting a chemical reaction which can limit shelf-life, such as the control of rancidity or oxidative discoloration by antioxidants. Neither does it include a number of food additives which are used primarily for other purposes but have been shown to contribute some antimicrobial activity. These include the antioxidants, butylatedhydroxytoluene (BHT) and butylatedhydroxyanisole (BHA), and the phosphates used as acidity regulators and emulsifiers in some products.

Preservatives may be microbicidal and kill the target organisms or they may be microbistatic in which case they simply prevent them growing. This is very often a dose-dependent feature; higher levels of an antimicrobial proving lethal while the lower concentrations that are generally permitted in foods tend to be microbistatic. For this reason chemical preservatives are useful only in controlling low levels of contamination and are not a substitute for good hygiene practices.

Recently consumers have shown an inclination to regard preservatives as in some way 'unnatural', even though the use of salts, acid, or smoke to preserve foods goes back to the beginning of civilization. Usage of chemical preservatives is now more restricted and controlled than ever and in many areas it is declining.

It is perhaps well to remember though that only the fairly recent advent of technologies such as canning and refrigeration has allowed us any alternative to chemical preservation or drying as a means of extending the food supply.

Natural Food Preservatives

The uncertainty voiced by consumer organizations and pressure groups over the use of food additives including preservatives has already been referred to. One approach to reassuring the consumer has been recourse to methods of preservation that can be described as 'natural'.

The whole area though is riddled with inconsistency and contradiction; it can be argued that any form of preservation which prevents or delays the recycling of the elements in plant and animal materials is unnatural.

On the other hand there is nothing more natural than strychnine or botulinum toxin. Smoking of foods might be viewed as a natural method of preservation. Its antimicrobial effect is a result of drying and the activity of wood smoke components such as phenols and formaldehyde which would probably not be allowed were they to be proposed as chemical preservatives in their own right.

The use of natural food components possessing antimicrobial activity such as essential oils and the lacto-peroxidase system in milk have attracted some attention in this respect. Attention has also been paid to the bacteriocins produced by food-grade micro-organisms such as the lactic acid bacteria. Nisin is an already well-established example and its use can be extended by expedients such as inclusion of whey fermented by a nisin-producing strain of Lactococcuslactis as an ingredient in formulated products like prepared sauces.

Modification of Atmosphere

At the start of the 19th Century it was believed that contact with air caused putrefaction and that food preservation techniques worked by excluding air. We have already seen how this misapprehension applied in the early days of canning and it was thought that drying operated in a similar way, expelling air from the interior of food.

Some preservation techniques, such as covering a product with a melted fat and allowing it to set, did in fact rely on the exclusion of air but it is only in the last 30 years or so that shelf-life extension techniques based on changing the gaseous environment of a food have really come to be widely used.

Modified atmospheres exert their effect principally through the inhibition of fast-growing aerobes that would otherwise quickly spoil perishable products. Obligate and facultative anaerobes such as Clostridia and the Enterobacteriaceae are less affected. Thus keeping quality is improved but there is generally little effect on pathogens, if present, and the technique is invariably applied in conjunction with refrigerated storage.

In practice three different procedures are used to modify the atmosphere surrounding a product: vacuum packing, modified-atmosphere packing or gas flushing, and controlled atmospheres. An essential feature of all three techniques is that the product is packed in a material which helps exclude atmospheric oxygen and retain moisture. This requires that it should have good barrier properties towards oxygen and water and be easily sealed.

The packaging materials used are usually plastic laminates in which the innermost layer is a plastic such as polyethylene which has good heat sealing properties. Mechanical closures on packs are far less effective as they often leave channels through which high rates of gas exchange can occur.

Overlying the layer of polythene is usually another layer with much better gas barrier properties. No plastics are completely impermeable to gases, although the extent of gas transmission across a plastic film will depend on the type of plastic, its temperature, the film thickness and the partial pressure difference across the film.

In some cases, it can also be affected by factors such as humidity and the presence of fat. Polyvinylidene chloride, PVDC, is a material commonly used as a gas barrier; the oxygen permeability of a 25/μm thick film is 10 $cm^3\ m^{-2}(24h)^{-1}atm^{-1}$ compared with values of 8500 and 1840 for low density and high density polythene respectively. Higher rates of transfer occur with CO_2 for which the permeability values are about five times those for oxygen.

If a film is required to exclude oxygen transfer completely, then a non-plastic material such as aluminium foil must be included. This is seen for example in the bags used to pack wines. In addition to the sealing- and gas barrier-layers, laminates may also contain an outer layer such as nylon which gives the pack greater resistance to damage.

In vacuum packing the product is placed in a bag from which the air is evacuated, causing the bag to collapse around the product before it is sealed. Residual oxygen in the pack is absorbed through chemical reactions with components in the product and any residual respiratory activity in the product and its microflora.

To achieve the best results, it is important that the material to be packed has a shape that allows the packaging film to collapse on to the product surface entirely – without pockets and without the product puncturing the film.

Vacuum packing has been used for some years for primal cuts of red meats. At chill temperatures, good quality meat in a vacuum pack will keep up to five times longer than aerobically stored meats. The aerobic microflora normally associated with the spoilage of conventionally stored meats is prevented from growing by the high levels of CO_2 which develop in the pack after sealing and the low oxygen tension.

The microflora that develops is dominated by lactic acid bacteria which are metabolically less versatile than the Gram-negative aerobes, grow more slowly and reach a lower ultimate population.

In recent years vacuum packing has been increasingly used for retail packs of products such as cooked meats, fish and prepared salads. It has been used less often for retail packs of red meats since the meat acquires the purple colour of myoglobin in its un-oxygenated form.

This does not appeal to consumers even though oxygenation occurs very rapidly on opening a vacuum pack and the meat assumes the more familiar bright red, fresh meat appearance of oxymyoglobin. Cured meats, on the other hand, are often vacuum packed for display since the cured meat pigment nitrosomyoglobin is protected from oxidation by vacuum packing.

The expanding range of chilled foods stored under vacuum and the availability of vacuum packing equipment for small-scale catering and domestic use has prompted concern about increasing the risk from psychrotrophic Clostridium botulinum.

A number of surveys have been conducted to determine the natural incidence of C. botulinum in these products and the concensus is that it is very low. In one recent example, workers failed to isolate C. botulinum or detect toxin in more than 500 samples analysed.

When they deliberately inoculated these products with C. botulinum spores and incubated at the abuse temperature of 10 °C, only in the case of vacuum packed whole trout was toxin produced within the declared shelf-life of the product.

Nevertheless, misuse of the technique does have the potential for increasing risk and a Government committee has recommended that all manufacturers of vacuum packing machinery should include instructions alerting the user to the risks from organisms such as C. botulinum.

In a variant of vacuum packing, known as cuisine sous-vide processing, food is vacuum packed before being given a pasteurization treatment which gives it a longer shelf-life under chill storage. The technique was developed in the 1970s in France and is said to give an improved flavour, aroma and appearance. It is used for the manufacture of chilled ready meals for various branches of the catering industry and sous-vide meals are also available in the retail market in some European countries.

They have been slow to appear on the UK market due to the lack of appropriate UK regulations and concern over their microbiological safety with respect to psychrotrophic C. botulinum. It has been recommended that sous-vide products with an intended shelf-life of longer than 10 days at <3 °C should receive a minimum heat process equivalent to 90 °C for 10 minutes; 70 °C for 100 minutes, should be sufficient for products with shorter shelf-lives.

In modified atmosphere packing, MAP, a bulk or retail pack is flushed through with a gas mixture usually containing some combination of carbon dioxide, oxygen and nitrogen. The composition of the gas atmosphere changes during storage as a result of product and microbial respiration, dissolution of CO_2 into the aqueous phase, and the different rates of gas exchange across the packing membrane. These changes can be reduced by increasing the ratio of pack volume to product mass although this is not often practicable for other reasons.

The initial gas composition is chosen so that the changes which occur do not have a profound effect on product stability. Some examples of MAP gas mixtures used in different products are presented in table. Carbon dioxide is included for its inhibitory effect, nitrogen is non-inhibitory but has low water solubility and can therefore prevent pack collapse when,high concentrations of CO_2 are used. By displacing oxygen it can also delay the development of oxidative rancidity. Oxygen is included in gas flush mixtures for the retail display of red meats to maintain the bright red appearance of oxymyoglobin.

MAP Gas Mixtures used with Foods

Product	%CO_2	%O_2	%N_2
Fresh meat	30	30	40
	15-40	60-85	-
Cured meat	20-50	0	50-80
Sliced Cooked roast beef	75	10	15
Eggs	20	0	80
	0	0	100
Poultry	25-30	0	70-75
	60-75	5-10	>20
	100	0	0

	20-40	60-80	0
Pork	20	80	0
Processed meats	0	0	100
Fish (white)	40	30	30
Fish (oily)	40	0	60
	60	0	40
Cheese (hard)	0-70		3-100
Cheese	0	0	100
Cheese grated/ sliced	30	0	70
Sandwiches	20-100	0-10	0-100
Pasta	0	0	100
	70-80	0	20-30
Bakery	0	0	100
	100	0	0

This avoids the acceptability problem associated with vacuum packs of red meats, although the high oxygen concentration (typically 60-80%) helps offset the inhibitory effect of the CO_2 (around 30%) so that the growth of aerobes is slowed rather than suppressed entirely.

In controlled-atmosphere storage, CAP, the product environment is maintained constant throughout storage. It is used mainly for bulk storage and transport of foods particularly fruits and vegetables, such as the hard cabbages used for coleslaw manufacture. CAP is used for shipment of chilled lamb carcasses and primal cuts which are packed in an aluminium foil laminate bag under an atmosphere of 100% CO_2.

It is more commonly encountered though with fruits such as apples and pears which are often stored at sub-ambient temperatures in atmospheres containing around 10% CO_2. This has the effect of retarding mould spoilage of the product through a combination of the inhibitory effect of CO_2 on moulds and its ability to act as an antagonist to ethylene, delaying fruit senescence and thus maintaining its own ability to resist fungal infection.

Control of Water Activity

The water activity of a product can be reduced by physical removal of liquid water either as vapour in drying, or as a solid during freezing. It is also lowered by the addition of solutes such as salt and sugar.

The primal role of these techniques in food preservation has been alluded to in a number of places. It was the earliest food preservation technique and, until the 19th Century, water activity reduction played some part in almost all the known procedures for food preservation.

Nature provided early humans with an object lesson in the preservative value of high solute concentrations in the form of honey produced by bees from the nectar of plants. The role of salt in decreasing a_w accounts for its extreme importance in the ancient economy as evidenced today in the etymology of the word, salary, and of place names such as Salzburg, Nantwich, Moselle and Malaga.

It can also be seen in the extraordinary hardship people were prepared to endure (or inflict on others!) to ensure its availability; to this day the salt mine remains a by-word for arduous and uncomfortable labour.

Solar drying while perhaps easy and cheap is, in many areas, subject to the vagaries of climate. Drying indoors over a fire was one way to avoid this problem and one which had the incidental effect t of imparting a smoked flavour to the food as well as the preservative effect of chemical components of the smoke.

Salting and drying in combination have played a central role in the human diet until very recently. One instance of this is the access it gave the population of Europe to the huge catches of cod available off Newfoundland. From the end of the 15th Century, salted dried cod was an important item in trans-Atlantic trade and up until the 18th Century accounted for 60% of all the fish eaten in Europe.

It remains popular today in Portugal and in the Caribbean islands where it was originally imported to feed the slave population. Other traditional dried and salted products persist in the modern diet such as dried hams and hard dry cheeses but the more recent development and application of techniques such as refrigeration, MAP, and heat processing and the preference for 'fresh' foods has meant that their popularity has declined. Nevertheless this should not obscure the important role that low a_w foods still play in our diet in the form of grains, pulses, jams, bakery products, dried pasta, dried milk, instant snacks, desserts, soups, etc.

Among the main features of the effect of a_w on the growth and survival of microorganisms it was noted that microbial growth does not occur below an a_w of 0.6. This applies to a number of food products, but the fact that microbial spoilage is not possible given proper storage conditions, does not mean that they do not pose any microbiological problems. Microorganisms that were in the product before drying or were introduced during processing can survive for extended periods.

This is most important with respect to pathogens if they were present in hazardous numbers before drying or if time and temperature allow them to resume growth in a product that is rehydrated before consumption.

There have been a number of instances where the survival of .pathogens or their toxins has caused problems in products such as chocolate, pasta, dried milk and eggs. Generally Salmonella and Staphylococcus aureus have been the principal pathogens involved – there have been 17 major outbreaks associated with these organisms and dried milk in the last 40 years, but spore formers are particularly associated with some other dried products such as herbs or rice.

Intermediate moisture foods, IMFs, are commonly defined as those foods with an a_w between 0.85 and 0.6. This range, which corresponds roughly to a moisture content of 15-50%, prohibits the growth of Gram-negative bacteria as well as a large number of Gram-positives, yeasts and moulds, giving the products an extended shelf-life at ambient temperature.

When spoilage does occur, it is often a result of incorrect storage in a high relative humidity environment. In correctly stored products growth of xerophilicmoulds, osmophilic yeasts or halophilic bacteria may occur, depending on the product, and in many IMFs the shelf-life is further protected by the inclusion of antifungal agents "such as sulfur dioxide or sorbic acid.

At the a_w of IMFs, pathogens are also prevented from growing. .Although Staph, aureus is capable of growing down to an a_w of 0.83, it cannot produce toxin and is often effectively inhibited by the combination of a_w with other antimicrobial hurdles.

There are a number of traditional IMFs such as dried fruits, cakes, jams, fish sauce and some fermented meats. Sweetened condensed milk is one interesting example. Milk is homogenized, heated to 80 °C and sugar added before it is concentrated in a multi-effect vacuum evaporator at 50-60 C. When the product emerges from the concentration stage it is cooled and seeded with lactose crystals to induce crystallization of the lactose.

This gives sweetened condensed milk its characteristic gritty texture. Although the product is packed into cans and has an almost indefinite shelf-life, it is not an appertized food. Its stability is a result of its high sugar content (62.5% in the aqueous phase) and low (<0.86). Spoilage may sometimes occur due to growth of osmophilic yeasts or, if the can is under-filled leaving a headspace, species of Aspergillus or Penicillium may develop on the surface.

Some years ago, our developing understanding of the stability of IMFs led to considerable interest in applying the same principles to the development of new shelf stable foods: Novel humectants such as glycerol, sorbitol and propylene glycol were often used to adjust a_w in these products in addition to the solutes salt and sugar.

They were not however well received in the market for human food because of acceptability problems, although a number of successful pet food products were developed. One interesting observation made during this work is that products with the same water activity differ in their keeping quality depending on how they are made. Traditional IMFs are generally made by a process of desorption whereby water is lost from the product during processing but a number of the new IMFs used an adsorption process in which the product is first dried and its moisture content readjusted to give the desired a_w.

The hysteresis effect in water sorption isotherms means that although products made using the two techniques will have the same initial a_w they will have different moisture contents and so will eventually equilibrate to different a_w values. It was found that products made by desorption and having the higher water content were also more susceptible to microbial spoilage.

Solar drying is still widely practiced in hot climates for products such as fruits, fish, coffee and grain. The traditional technique of spreading the product out in the sun with occasional turning often gives only rudimentary or, sometimes, no protection from contamination by birds, rodents, insect and dust.

Rapid drying is essential to halt incipient spoilage; this is usually achievable in hot dry climates, though in tropical countries with high humidity drying is usually slower so that products such as fish are often pre-salted to inhibit microbial growth during drying.

There are a number of procedures for mechanical drying which are quicker, more reliable, albeit more expensive than solar drying. The drying regime must be as rapid as possible commensurate with a high-quality product so factors such as reconstitution quality must also be taken into account. With the exception of freeze- drying where the product is frozen and moisture sublimed from the product under vacuum, these techniques employ high temperatures. During drying a proportion of the microbial population will be killed and sub-lethally injured to an extent which depends on the drying technique and the temperature regime used.

It is however no substitute for bactericidal treatments such as pasteurization. Although the air temperature employed in a drier may be very high, the temperature experienced by the organisms in the wet product is reduced due to evaporative cooling. As drying proceeds and the product temperature increases, so too does the heat resistance of the organisms due to the low water content.

This can be seen for example in the differences between spray dried milk and drum dried milk. In spray drying, the milk is pre-concentrated to about 40-45% solids before being sprayed into a stream of air heated to temperatures up to 260 °C at the top of a tower.

The droplets dry very rapidly and fall to the base of the tower where they are collected. In drum drying, the milk is spread on the surface of slowly rotating metal drums which are heated inside by steam to a temperature of about 150 °C.

The film dries as the drum rotates and is scraped off as a continuous sheet by a fixed blade close to the surface of the drum. Although it uses a lower temperature, drum drying gives greater lethality since the milk is not subject to the same degree of pre-concentration used with spray-dried milk and the product spends longer at high temperatures in a wet state.

Spray drying is however now widely used for milk drying because it produces a whiter product which is easier to reconstitute and has less of a cooked flavour. Milk is pasteurized before drying although there are opportunities for contamination during intervening stages. Most of the organisms which survive drying are thermoduric but Gram-negatives may survive and have on occasion been the cause of food poisoning outbreaks.

The limited lethality of drying processes and the long storage life of dried products means that manufacturers are not exempt from the stringent hygiene requirements of other aspects of food processing. Good quality raw materials and hygienic handling prior to drying are essential.

Outbreaks of Staph, aureus food poisoning have been caused by dried foods which were stored at growth temperatures for too long prior to drying allowing the production of heat resistant toxin which persisted through to the final product. The dried product must also be protected from moisture by correct packaging and storage in a suitable environment otherwise pockets of relatively high a_w may be created where microbial growth can occur.

Compartmentalization

Butter is an interesting example of a rather special form of food preservation where microbial growth is limited by compartmentalization within the product. Essentially there are two types of butter: sweet cream butters, which are often salted, and ripened cream butters. In ripened-cream butters, the cream has been fermented by lactic acid bacteria to produce inter alia acetoin from the fermentation of citrate which gives a characteristically buttery flavour to the product.

They have a stronger flavour than sweet-cream butters but are subject to faster chemical deterioration. Sweet-cream butter is most popular in the United States, Ireland, the UK, Australia and New Zealand whereas the ripened cream variety is more popular in continental Europe.

Butter is an emulsion of water droplets in a continuous fat phase in contrast to milk which is an emulsion of fat globules in a continuous water phase. It has a higher fat content than milk (80%) and uses pasteurized cream as its starting point.

Typically, the cream is pasteurized using an HTST process of 85 °C for 15s and held at 4-5 °C for a period to allow the fat globules to harden and cluster together. In making a conventional ripened cream butter, the starter culture is added at this stage and the cream incubated at around 20 °C to allow flavour production to take place.

A more recent method developed at NIZO, the Dutch Dairy Research Institute, employs a concentrated starter added to sweet-cream butter after manufacture. Phase inversion, the conversion from a fat-in-water emulsion to a water-in-fat emulsion, is achieved by the process of churning.

During this process fat globules coalesce, granules of butter separate out, and considerable amounts of water are lost from the product in the form of buttermilk. The buttermilk phase retains most of the micro-organisms from the cream and numbers may show an apparent increase due to the breaking of bacterial clumps.

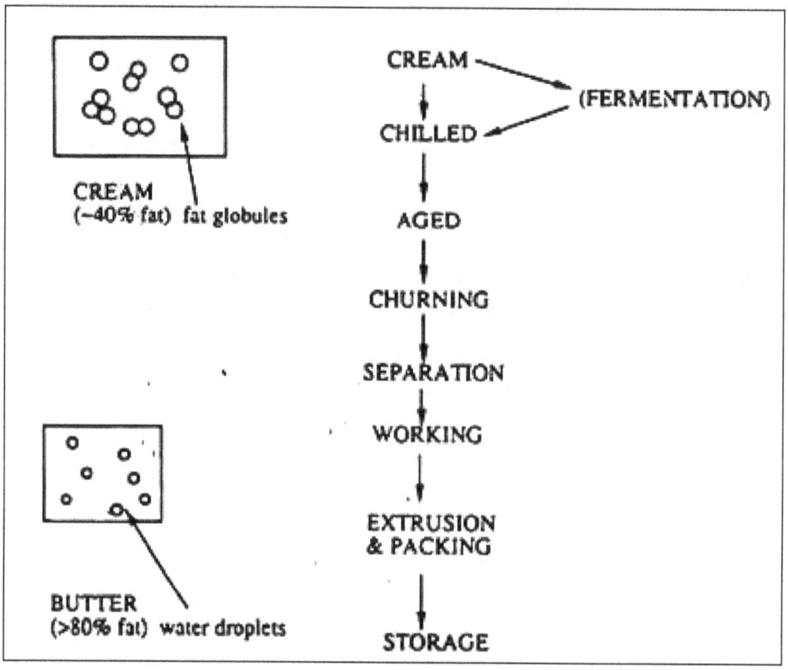

Buttermaking.

Traditional farmhouse butter making used wooden butter churns and these were originally scaled up for the earliest commercial butter making. However the impossibility of effectively cleaning and sanitizing wood has led to its replacement by churns made of stainless steel or aluminium-magnesium alloys. After the butter has formed, the buttermilk is drained off, the butter grains washed with water and, in the case of sweet-cream butter, salt is added usually at a level of 1-2%.

The butter is then 'worked' to ensure further removal of moisture and an even distribution of water and salt throughout the fat phase. In properly produced butter the water is distributed as numerous droplets ($>10^{10}$ g^{-1}) mostly less than 10/μm in diameter. Since the butter should contain at most around 10^3cfu g^{-1} most of these droplets will be sterile.

In those that do contain micro-organisms, the nutrient supply will be severely limited by the size of the droplet. If the butter is salted, the salt will concentrate in the aqueous phase along with the bacteria which will therefore experience a higher, more inhibitory salt level. For example, bacteria in a butter containing 1% salt and with a moisture content of 16% would experience an effective salt concentration of 6.25%.

Few micro-organisms survive pasteurization so the microbiological quality of butter depends primarily on the hygienic conditions during subsequent processing, particularly the quality of the water used to wash the butter. Good microbiological quality starting materials are essential though, as preformed lipases can survive pasteurization and rapidly spoil the product during storage.

Butter spoilage is most often due to the development of chemical rancidity but microbiological problems do also occur in the form of cheesy, putrid or fruity odours or thc rancid flavour of butyric acid produced by butterfat hydrolysis. Pseudomonads are the most frequently implicated cause and are thought to be introduced mainly in the wash water.

Psychrotrophic yeasts and moulds can also cause lipolytic spoilage and these are best controlled by maintaining low humidity and good air quality in the production environment and by ensuring the good hygienic quality of packaging materials. In this respect aluminium foil wrappers are preferred to oxygen- permeable parchment wrappers as they will help discourage surface mould growth.

Margarine relies on a similar compartmentalization for its microbiological stability, but uses vegetable fat as its continuous phase. Although skim milk is often included in the formulation, it is possible to make the aqueous phase in margarine even more deficient nutritionally than in butter, thus increasing the microbiological stability further.

With the move towards low fat spreads containing 40% fat, the efficacy of this system is more like y to breakdown. A higher moisture content means that the preservative effect of salt or lactic acid, which is often included, is diluted and that micro-organisms can grow to a greater extent in the larger aqueous droplets. In these cases the use of preservatives may be required to maintain stability.

An approach developed is based on a two stage approach where the composition of the aqueous phase is analysed to determine its capacity to support the growth of different spoilage organisms.

If there is some potential for microbial growth to occur, this is then calculated by working out which fraction of water droplets will be contaminated and then summing the growth in each of them. These models have been incorporated into an expert system to predict the stability of any proposed product formulation so that microbiological stability can be designed into the product.

References

- Food-preservation, sports-and-everyday-life-food-and-drink-food-and-cooking: encyclopedia.com, Retrieved 23 April, 2020
- Rickman (2007). "Nutritional comparison of fresh, frozen, and canned fruits and vegetables". Journal of the Science of Food and Agriculture. 87 (7): 1185–1196. doi:10.1002/jsfa.2824
- Cerf, O.; Condron, R. (2006). "Coxiella burnetii and milk pasteurization: an early application of the precautionary principle?". Epidemiology & Infection. 134 (5): 946–51. doi:10.1017/S0950268806005978. ISSN 1469-4409. PMC 2870484. PMID 16492321

- Si X, Chen Q, Bi J, Wu X, Yi J, Zhou L, Li Z (2016). "Comparison of different drying methods on the physical properties, bioactive compounds and antioxidant activity of raspberry powders". J Sci Food Agric. 96 (6): 2055–62. doi:10.1002/jsfa.7317. PMID 26108354
- García, Pilar; Rodríguez, Lorena; Rodríguez, Ana; Martínez, Beatriz (2010). "Food biopreservation: promising strategies using bacteriocins, bacteriophages and endolysins". Trends in Food Science & Technology. 21 (8): 373–382. doi:10.1016/j.tifs.2010.04.010. hdl:10261/51440
- Methods-of-food-preservation-top-10-methods, foods-preservation-foods-59270: biologydiscussion.com, Retrieved 15 May, 2020

4

Food Safety

The practices which aim at preserving the quality of food and prevent it from contamination and food borne illness during preparation, handling and storage are referred to as food safety. Safe food handling, frozen foods, food hazard analysis, food packaging, etc. are some of its concepts. The topics elaborated in this chapter will help in gaining a better perspective of food safety.

Food can become contaminated at any stage of production, processing, distribution, storage or preparation. For example, germs can spread to food from unclean surfaces, utensils or equipment are used whether during food production or at home in our own kitchen. If chilled raw foods like meat or dairy products are left at room temperature for too long, for example during transport to or from the supermarket, bacteria can grow faster than expected and pose a safety risk.

Food safety is about handling, storing and preparing food to prevent infection and help to make sure that our food keeps enough nutrients for us to have a healthy diet.

The term food safety describes all practices that are used to keep our food safe. Food safety relies on the joint efforts of everyone involved in our food supply. All along the food chain, from farmers and producers to retailers and caterers, legislation and controls are in place to reduce the risk of contamination, and personally we each have a role to play as well.

Many people get sick from their food at home, and bacteria (such as Campylobacter and Salmonella) and viruses (such as norovirus) are common causes. Unsafe food and water means that it has been exposed to dirt and germs, or may even be rotten, which can cause infections or diseases such as diarrhoea, meningitis, etc. These diseases can make people very sick or even be life threatening. When people are sick, they are weak and would have difficulty working or concentrating at school. Some of these infections also make it difficult for our bodies to absorb the nutrients they need to get healthy. Unsafe or stale foods also deteriorate and are of poor quality, which means they lose nutrients and so we do not get enough of what we need for a healthy diet. So unsafe food can also lead to poor nutrition.

Sticking to some basic safe food handling tips can help us avoid getting sick.

Keep Clean

Microbes spread to food through physical contact, for example from your hands or cooking utensils. Always wash your hands thoroughly with warm soapy water before handling food, and repeat often while cooking. Prepare and chop food on a clean surface and clean all utensils and surfaces thoroughly after use with hot water and detergent, or in the dishwasher. Wash dishcloths, tea towels, and aprons frequently at high temperatures.

Separate Raw and Cooked

Raw foods like meat, fish and poultry are most likely to contain illness-causing microbes. These can be transferred to ready-to-eat foods by cross-contamination directly (for example if raw meat touches cooked foods) or indirectly (for example chopping salad vegetables with a knife that was used to chop raw meat). To avoid cross-contamination, try dedicating different coloured chopping boards to fruit/vegetables, fish, meat/poultry and cooked foods, so you always know which to use. Use separate re-usable shopping bags for raw and ready-to-eat foods and label them so you remember which is which.

Cook Thoroughly

Cooking/heating foods to temperatures until piping hot throughout will kill most illness-causing microbes. The most reliable way to check the temperature is to use a cooking thermometer - check that your food has reached a core temperature of least 72 °C for 2 minutes.

- Whole cuts of beef or lamb can be eaten rare or pink as they are unlikely to have harmful bacteria in the centre. The outer surface should be seared.
- For pork and poultry, there should be no pink meat left. If you don't have a thermometer, pierce the thickest part with a fork or skewer; the juices should run clear, not pink.
- Ground meat/fish products like burgers, sausages or fishcakes should be cooked thoroughly all the way through.
- Reheat leftovers thoroughly all the way through. Bring soups and stews to the boil for at least 2 minutes.

Store at Safe Temperature

Not all foods need to be refrigerated. Clean, dry and cool shelves are the best place to store bread, dry food (in sealed bags or containers), unopened tins and jars. Foods like milk, meat, fish, poultry, and also our own leftovers should be kept in the fridge, to slow down the growth of harmful microbes.

- Remember the '2-hour rule': whether it is leftover, take-away, or just-bought, cool and place foods in the fridge within 2 hours.

- Keep the fridge below 5 °C - check regularly with a fridge thermometer.
- Store raw meat, fish and poultry in sealed containers, on the bottom shelf of your fridge to avoid spread of germs to ready-to-eat foods.
- When transporting food (e.g. taking your lunch to work) use an insulated bag or cold box with ice pack (or a carton of frozen juice!) to keep your meal at a safe temperature.

The world's growing population and the consumers' desire to be provided with a wider range of foods have resulted in a longer and more complex food chain. Today, foods reach consumers after being collected from fields, farms and factories and then pass onto many countries, traveling distances of thousands of kilometers. With this global food distribution, an infection that occurs at any point within the food chain has the potential of affecting any given population in the world. It is therefore essential, given the number of interactions taking place between the actors involved in the food chain and the long distances between them, that multi-sectorial and international collaboration take place. As no country can provide food safety on its own, safety measures need to be enhanced in many countries.

While experts on food safety and health have determined that millions of foodborne disease cases are reported every year, the actual numbers are clouded by uncertainty, as most cases go unreported. Furthermore, foodborne diseases are difficult to diagnose, since they have various symptoms, including fatigue, chills, mild fever, vertigo, upset stomach, dehydration caused by diarrhea, severe cramps and, in some cases, even death. In many of the reported cases, foods prepared outside of the home are the primary cause of foodborne diseases, though it is not uncommon for home-made foods to also cause diseases. Studies conducted on the distribution of foodborne diseases across the world have demonstrated that a majority of these diseases occur during the processing of the food in the preparation stage at home or at food production sites. In fact, most foodborne diseases can be prevented if the regulations governing food safety were complied with, from production stages to consumption.

Improper heating of the food, such as undercooking, re-heating and waiting in the heat, or improper cooling of the food account for 44% of the foodborne illnesses. Inadequate preparation and improper cooking practices, such as those involving cross-contamination, insufficient processing, poor hygiene and the re-use of leftovers, are responsible for causing 14% of these diseases.

Foodborne diseases are widespread throughout the world. The process by which a foodborne disease spreads begins with the features of the disease contaminating the food, which in turn threaten both individual and public health by means of the foods. Healthy, or what can be termed as safe, food is food that has not lost its nutritional val-ue, that is clean, in physical, chemical and microbiological terms and that is not stale. The factors causing the contamination of the food may threaten the safe consumption of it and thereby make the foods harmful to human health. For this reason, it is necessary to utilize various resources to prevent the food from being contaminated at all stages of the food chain, from harvest to consumption.

Food Safety Systems

Effective food control systems are needed to improve the applicability and control of food safety. Currently, the HACCP, ISO 22000 and PAS 220 are the most commonly used internationally approved food safety systems.

Hazard Analysis and Critical Control Points (HACCP)

HACCP was first used in the 1960s by the American Pillsbury company for the purpose of producing "zero defect" products for the US Army and NASA. Later, starting in the 1970s, it began to be used as a reference by the Food and Drug Administration (FDA) in official supervisions. It was adopted by the Codex Alimentarius Commission in 1992 and published as the HACCP international standard for the first time. Since then, the food industry and official authorities have been using it to protect against and control the risks of potential dangers that could threaten food safety.

Initially, HAACCP had three principles:

- Identification and assessment of hazards associated with food products.
- Determination of critical control points to control identified hazards.
- Establishment of a system to monitor the critical control points.

The HACCP, as it is applied today, has five starting steps and is governed by seven principles. The starting steps were created by Codex, and they should be completed prior to implementing the seven HACCP principles. The starting steps help to ensure that the HACCP system is implemented and managed in the most effective way possible.

The HAACP system is applicable for any company operating within the food chain, regardless of their size. In the implementation stage, the HACCP system should be supported by certain preliminary condition programs. A company interested in implementing this system should already be following the requirements of this preliminary condition program. Preliminary condition programs include national regulations, codes of practice or other food safety prerequisites. In general, preliminary condition programs involve factories and equipment, staff training, cleaning and sanitation, maintenance chemical control, waste management, storage and transportation.

HACCP Implementation in 12 Steps

In 2005, The International Organization for Standardization (ISO) published a standard for the Food Safety Management System known as ISO 22000. The ISO 22000 system is a combination of preliminary condition programs, HACCP principles and implementation steps defined by the Codex Alimentarius Commission and ISO 9001:2000 standard components. After it was defined, it began to be used in more than 50 countries within 2 years.

Five starting steps	The seven principles of the HACCP system
1. Assemble HACCP team	1. Conduct a hazard analysis
2. Describe the product	2. Determine Critical Control Points
3. Identify intended use	3. Establish critical limits for each CCP
4. Construct flow diagram	4. Establish a monitoring system for each CCP
5. Conduct on-site confirmation of flow diagram	5. Establish corrective actions
	6. Establish verification procedures
	7. Establish documentation and record keeping

The basic approach of the ISO 22000 standard is to implement a preventive system that serves to protect consumers from foodborne diseases. This standard controls all the processes in the food chain, including infrastructure, staff and equipment. In business establishments, the Food Safety Management System implementations include production control, product control, equipment control, maintenance, general hygiene practices, staff and visitor hygiene, transportation, storage, product information, training, the selection and evaluation of suppliers, communication and other similar issues.

The main goal of this standard is to have a system in place that determines the unacceptable risks that may result from process errors and to secure product safety and consumer health. Food safety supervision over product, design, production and quality control determines and eliminates the potential dangers. The fundamental role of ISO 22000 is not only to provide food safety but also to improve the sensory and nutritional quality of food, and it also plays a primary role in the quality assurance of service practices in industrial production. Lastly, this standard helps to reduce operational losses by instituting a more effective use of resources to increase productivity, and thereby, directs the establishment to a system of total quality.

PAS 220 (Publicly Available Specification)

This standard was created by the major global food producers in cooperation with the Confederation of Food and Drink Industries (CIAA) with the purpose of eliminating the weaknesses of the ISO 22000 food safety system standard. Nestle, Unilever, Danone and Kraft, the sector leaders generally known as "G4", collectively published the PAS 220 standard, which refines the preliminary conditions programs. The PAS 220 standard is applicable for all types of companies and was made available in 2008. It was intended that the PAS 220 standard be used together with the internationally accepted ISO 22000 standard.

The content and topics of PAS 220 elaborate on the 10 sub-titles in the ISO 22000 standard and adds 5 of its own, resulting in the following 15 items:

- Structure and placement of buildings,
- Placement of work site, buildings and their wings,
- Supporting plants (air, water, energy),
- Supporting services, including wastes and sewage,
- Adequacy of the equipment, cleaning and preventive care,
- Management of purchased materials,
- Measures against cross-contamination,
- Cleaning and sanitation,
- Pest control,
- Staff hygiene and workers' lodgings,
- Re-processing,
- Product recall procedures,
- Storage,
- Informing consumers about products,
- Food defense, biodefense and bioterrorism.

Good Agricultural Practices

Today, increasing attention is focused upon the impact farming practices are having on the environment, and there is an increasing emphasis on more sustainable methods of crop production. Systems need to be adopted that are more sensitive to environmental issues, genetic diversity, wildlife and their habitats and in some cases the social structures of rural communities. Furthermore, consumers around the world are more sophisticated and critical than in the past, demanding to know how and what has been used to produce their agriculturally derived products.

Good Agricultural Practices (GAP) are defined "practices that address environmental, economic and social sustainability for on-farm processes, and result in safe and quality food and non-food agricultural products" by the FAO. The aims of the GAP are as follows:

- Ensuring agricultural production harmless to environment, human and animal health,
- Safety of natural resources,

- Ensuring traceability and sustainability in agriculture,
- Improving workers health and working conditions,
- Ensuring safety and quality of produce in the food chain.

General principles for GAP were first presented to the FAO Committee on Agriculture (COAG) in 2003 in the paper "Development of a Framework for Good Agricultural Practices" the annex of which broadly outlined farm-level GAP recommendations in 10 fields; which are "soil", "water", "crop and fodder production", "crop protection", "animal protection", "animal health and welfare", "harvest and on-farm processing and storage", "energy and waste management", "human welfare, health, and safety" and "wildlife and landscape".

Although there are some GAP used by different organizations to succeed different purposes and goals, Globalgap (Eurepgap) is the widespread certificate in agricultural produce worldwide. Globalgap documents consists of; ISO 9001:2000 Quality Management System, ISO 14000 Environmental management system, OHSAS 18001 Work Health and Safety Management System and ISO 22000 Food Safety Management System principles.

Food safety ultimately deals with the consumption stage, where the existence and level of the dangers caused by foods are of chief concerns. The observance of rigorous control procedures throughout the course of the food chain is a fundamental necessity, given that risks to food safety can surface in any stage of the chain. Therefore, all parties involved in the food chain share the responsibility for ensuring food safety.

The design of a food safety system involves numerous factors. To begin with, minimum hygiene standards should be determined by laws and regulations, food producers must apply food safety measures and procedures and official bodies must supervise and inspect food industry companies to confirm that they are conducting their operations in a manner consistent with the regulations in force. Food poisoning cases that threaten public health globally occur as a result of the contamination of foods in any stage, from production to consumption. Although the factors jeopardizing food safety seem to be easy to control in theory, studies and current practices indicate that there is still a long way to go in practice.

Infant Food Safety

Foodborne illness (also foodborne disease and colloquially referred to as food poisoning) is any illness resulting from the food spoilage of contaminated food, pathogenic bacteria, viruses, or parasites that contaminate food. Infant food safety is the identification of risky food handling practices and the prevention of illness in infants. Fooddborne illness is a serious health issue, especially for babies and children. Infants and young children are particularly vulnerable to foodborne illness because their immune systems are not developed enough to fight off foodborne bacterial infections. In fact, 800,000 illnesses affect children under the age of 10 in the U.S. each year. Therefore, extra care should be taken when handling and preparing their food.

Prevention

Handwashing is the first step in maintaining the safety of infant food. Caregivers hands can pick up bacteria and spread bacteria to the baby. Objects that are high in bacteria are:

- Diapers containing feces and urine,
- Raw meat and raw poultry,
- Uncooked seafood, and eggs,
- Dogs and cats, turtles, snakes, birds, and lizards,
- Other animals,
- Soil,
- Other children.

Handwashing can remove harmful bacteria and will help to prevent foodborne illness. Instructing other children in a family on good handwashing will help to spread the bacteria that cause illness.

Handwashing is most effective in providing safe food for the infant during 'key times':

- Before preparing and feeding bottles or foods to the baby.
- Before touching the baby's mouth.
- Before touching pacifiers or other things that go into the baby's mouth.
- After using the toilet or changing diapers.

Infant Formula

Though breastfeeding helps prevent many kinds of sicknesses among infants, caregivers often choose to use infant formula. Promoting food safety in infants requires safe preparation and use.

Use infant formula within two hours of preparation. If the infant does not finish the entire bottle, the remainder is thrown away. If the prepared feed is not used right away, refrigerating it immediately will slow the growth of microorganisms, however it must be used within 24 hours.

Cronobacter, formerly known as Enterobacter sakazakii, is a group of bacteria that can be found in the environment. The germs can also live in dry foods, such as powdered infant formula. Anybody can get sick from *Cronobacter*, but the infection occurs most often in infants. *Cronobacter* infections are rare, but they can be deadly in newborns. Infections in infants usually occur in the first days or weeks of life. *Cronobacter* germs can cause dangerous blood infections (sepsis) or infections of the linings surrounding the brain and spine (meningitis). Infants two months of age and younger are most likely to develop meningitis if they get sick with Cronobacter. Infants born prematurely and infants with a lower ability to fight germs and sickness due to illness (such as HIV) or medical treatment (such as chemotherapy for cancer) are also more likely to get sick. The first symptom of *Cronobacter* infection in infants is usually a fever, coupled with poor feeding, crying, or very low energy. Parents or caregivers should take an infant with these symptoms to see a doctor.

Infection Prevention

Cronobacter infections can be prevented by:

- Breastfeeding. This is one of the best things that can do for the infant's health, and benefits include preventing many kinds of infections. Reports *Cronobacter* infections among infants who were fed only breast milk and no formula or other foods are rare.
- Cleaning, sanitizing, and storing dry feeding items and breast pump parts. Contamination can be prevented and the milk kept relatively free of microbes by washing, sanitizing, and safely storing other feeding implements.
- Cleaning the breasts before nursing.
- Keeping ill family and friends from feeding the infant.
- Using liquid formula, when possible. If your baby gets formula,
- Choosing infant formula sold in liquid form over powdered, especially in very young infants.
- Preparing powdered infant formula safely.

Safe Infant Formula Preparation

Recommendations from health organizations for the preparation of infant formula are:

- Warming treated and clean water to at least 158 °F /70 °C and pour it into the bottle.
- Adding infant formula, and carefully shaking, rather than stirring the bottle.
- Cooling the mixture so that it is not too hot before feeding the baby by running the prepared, capped bottle under treated, clean and cool water taking care to keep the cooling water from getting into the bottle or on the nipple.
- Testing the temperature of the mixture by shaking a few drops on the wrist.

If soap and water are not available, use an alcohol-based hand sanitizer with at least 60% alcohol (check the product label to be sure). Hand sanitizer with at least 60% alcohol is effective in killing Cronobacter germs. But use soap and water as soon as possible afterward because hand sanitizer does not kill all types of germs and may not work as well if hands are visibly greasy or dirty.

Cronobacter can also cause diarrhea and urinary tract infections in people of all ages. The infection can be serious for older people and for people whose immune systems are weakened by other illnesses or conditions. They are also more likely to get sick.

Caregivers can prepare infant formula safely by:

- Preparing safe water for mixing: Bring tap water to a roiling boil and boil it for one minute. Bottled water can be sterilized before using.
- Using clean bottles and nipples: Sterilizing bottles and nipples before first use is safer. After that, it is safe to wash them by hand or in a dishwasher.
- Not making more formula than is needed. Formula can become contaminated during preparation, and bacteria can multiply quickly if formula is improperly stored. The safest practice is to make formula in smaller quantities on an as-needed basis to greatly reduce the possibility of contamination and always follow the label instructions for mixing formula.

Heating Breast Milk or Infant Formula

There are two ways to heat bottles with disposable inserts or hard plastic, and glass bottles. A bottle can be placed under hot, running tap water until the desired temperature is reached. This should take one-to-two minutes. A bottle can be placed in a pan after the water has been heated on a stove. The pan can be removed from the heat and set the bottle in it until it is warm. It is safer to shake milk or formula to even out the temperature. Heating breast milk or infant formula in the microwave is not recommended. This results in "hot spots" that can scald a baby's mouth and throat.

Cow's Milk

Cow's milk by itself is not appropriate for infants less than one year old. Cow's milk does not have the correct balance of nutrients for infants to grow and develop normally, and it can cause problems with anemia and kidney function. Raw milk is never appropriate for infants – or anyone else. It should not be consumed by anyone at any time for any purpose. Raw milk can harbor dangerous microorganisms, such as *Salmonella*, *E. coli*, and *Listeria*, that can pose serious health risks. Most infant formula is made with cow's milk, but it has been modified and supplemented with additional nutrients. As a result, the formula is more nutritious and easier for the baby to digest than cow's milk. Other formula options include soy-based formulas and hypoallergenic (or protein hydrolysate and amino acid-based) formulas. Special formulas are available for babies who are premature or have other health problems.

Sanitizing Objects

Infants put anything within reach into their mouths. It is also important to keep all objects that enter baby's mouths (such as pacifiers and teethers) clean. Though research into five-second rule has been done, results are inconclusive.

Solid Foods

Infants are introduced to solid foods at different ages. Breast milk alone is sufficient to support optimal growth and development for approximately the first six months after birth. For these very young infants water, juice, and other foods are generally unnecessary. Even when babies enjoy discovering new tastes and textures, solid foods should not replace breastfeeding, but merely complement breast milk as the infant's main source of nutrients throughout the first year. Beyond one year, as the variety and volume of solid foods gradually increase, breast milk remains an ideal addition to the child's diet. Parents and caregivers can reduce choking hazards in a child's environment. Special attention should be given to food and nonfood items (e.g., candy, nuts, and coins) commonly involved in choking. Younger children are particularly at risk because of their tendency to place objects in their mouths, poor chewing ability, and narrow airways compared with those of older children. Recommendations are available to guide parents and caregivers about the types of food items that are inappropriate for children aged less than four years. Removal of nonfood choking hazards also is important for infants and children aged less than five years because approximately one third of all choking episodes involve nonfood items.

Microwaving of Solid Foods

When baby food is microwaved in a jar it often heats unevenly. The hottest places are in the center of the foods. The coolest places are next to the glass sides, which could lead you to believe that the food is not too hot. Safe practices for microwaving food for infants are:

- Don't microwave baby foods in the jar. Instead, transfer the food to a dish before microwaving it. This way the food can be stirred and taste-tested for temperature.
- Microwave four ounces of solid food in a dish for about 15 seconds on high power. Always stir, let stand 30 seconds, and taste-test before feeding. Food that's "baby-ready" should taste or feel lukewarm.
- Do not heat baby-food meats, meat sticks or eggs in the microwave. Use the stovetop instead. These foods have a high fat content, and since microwaves heat fats faster than other substances, these foods can cause splattering and overheating.

First Aid for Choking

Early and effective treatment is crucial to prevent morbidity and mortality from childhood choking. Methods taught routinely in courses on cardiopulmonary resuscitation (CPR) or first aid can be lifesaving when instituted early by trained parents and caregivers. Opening the airway quickly by ejecting the foreign body can avoid potentially severe injuries. The American Academy of Pediatrics recommends that all parents and caregivers participate in the American Heart Association's Basic Lifesaving Course or the American Red Cross' Infant/Child CPR Course. To prevent infant choking, be sure your child has adequate motor skills to swallow food. Do not offer babies or young children high-risk foods, such as chunks of meat, cheese, grapes, or raw vegetables, unless they are cut up into small pieces. Avoid hard foods, such as nuts, seeds, and popcorn.

Commercial Baby Food

Infant food safety includes the evaluation of commercially prepared baby food before feeding it to a baby. Examining each jar of commercial baby food before using it and noting the position of the safety button on top of the jar will indicate whether or not the jar has been opened. Unopened baby food in jars will have a safety button that is down. A jar of baby food that is swelling, leaking, contain chipped glass is probably not safe to feed to a baby. Some baby food comes in pouches. If the pouch is leaking or swelling it may not be safe to feed to a baby.

The Food and Drug administration has published the following *do's* and *don'ts* regarding commercial baby food safety:

- Do not "double dip" with baby food.
- Never put baby food in the refrigerator if the baby does not finish it.
- It is best if a baby is not fed directly from the jar of baby food.
- A small serving of food on a clean dish and refrigerating the remaining food in the jar is safer.
- If the baby needs more food from the same jar, using a clean spoon to serve another portion is safer.
- It is safer to throw away any food in the dish that is not eaten.
- If a baby is fed from a jar, it is safer to throw away any food left in the jar.
- Sharing spoons is not the safest way to feed a baby
- Putting the baby's spoon in the mouth of another person is not safe.
- Baby food left out for two hours or more will begin to grow bacteria at room temperature. It is not safe.

- If an opened baby food container is stored in the refrigerator for more than three days, it may not be safe to feed the baby. "If in doubt, throw it out".

Handwashing

- Wet hands thoroughly with warm water and add soap.
- Thoroughly scrub hands, wrists, fingernails, and in between fingers - for at least 20 seconds.
- Rinse, then dry hands with a clean cloth towel or use a paper towel so the germs are thrown away.

When to Wash

Handwashing prevents spreading infection. Washing before and after handling food, after using the bathroom, changing diapers, or handling pets.

FOOD SAFETY-RISK ANALYSIS

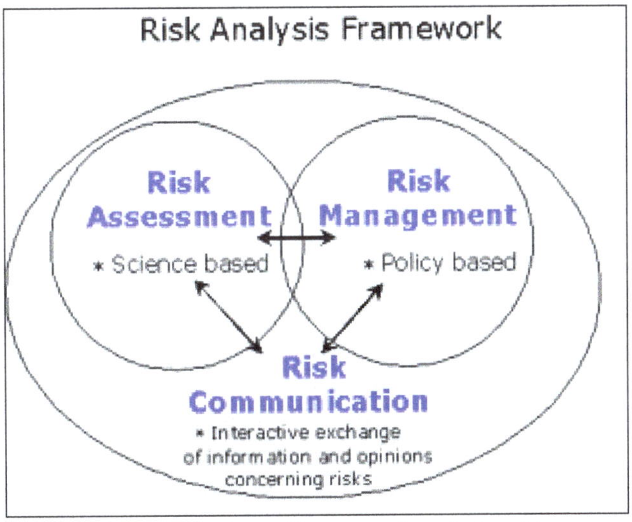

The diagram above illustrates the relationship between the three components of risk analysis.

A food safety-risk analysis is essential not only to produce or manufacture high quality goods and products to ensure safety and protect public health, but also to comply with international and national standards and market regulations. With risk analyses food safety systems can be strengthened and food-borne illnesses can be reduced. Food safety risk analyses focus on major safety concerns in manufacturing premises—not every safety issue requires a formal risk analysis. Sometimes, especially for complex or controversial analyses, regular staff is supported by independent consultants.

Risk Analysis

Risk analysis is defined for the purposes of the Codex Alimentarius Commission as "A process consisting of three components: risk management, risk assessment, and risk communication".

Risk Management

Risk management is defined for the purposes of the Codex Alimentarius Commission as "The process, distinct from risk assessment, of weighing policy alternatives, in consultation with all interested parties, considering risk assessment and other factors relevant for the health protection of consumers and for the promotion of fair trade practices, and, if needed, selecting appropriate prevention and control options."

Risk Assessment

General Characteristics

As defined by the Codex Alimentarius Commission and adopted by international food safety commissions, food safety risk assessment is "The scientific evaluation of known or potential adverse health effects resulting from human exposure to foodborne hazards." The most important aspect of risk assessment in relation to food safety is that it should be rooted in scientific data. Sources of data should be assembled in a systematic manner and should stem from valid scientific studies and communities across the world. A proper risk assessment can be described as being objective and unbiased, with absolute transparency. When at all possible, the assessment should remain independent of risk management as to preserve the integrity of the science and not have influence from regulatory policy and values. All assumptions made throughout the assessment should be well documented by the risk manager and should strive to be as objective, biologically realistic, and consistent as possible. As with any risk assessment performed, incomplete data or gaps in information create degrees of variability and uncertainty. In accounting for these factors, an extensive description of uncertainties in the risk estimate and their origins should be provided, as well as, descriptions of how assumptions being made can increase or decrease the uncertainty of results in the risk assessment. To increase the validity of a risk assessment, it is recommended that the assessment remain open for peer review and editing by food safety and science communities. A proper risk assessment is a constantly revolving process consisting of the following steps: (i) hazard identification, (ii) hazard characterization, (iii) exposure assessment, and (iv) risk characterization.

Hazard Identification

"The identification of biological, chemical, and physical agents capable of causing adverse health effects and which may be present in a particular food or group of foods."

This is often considered the most important step in a risk assessment as an unidentified hazard in the early stages of the production process can cause devastating effects in later stages.

- Potential biological hazards: bacteria, molds, yeasts, viruses, parasites, fish and shellfish as sources of toxic compounds, and pests (birds, insects, and rodents) as carriers of pathogens.
- Potential chemical hazards: toxic plant material, intentional/unintentional food additives, insecticides, pesticides, other agricultural chemicals, antibiotic or other drug residue, food allergens, food intolerances, excessive addition of nutrients, and anti-nutritional factors.
- Potential physical hazards: glass, wood, stones, metal, packaging materials, bones, and personal effects.

Hazard Characterization

"The qualitative and quantitative evaluation of the nature of the adverse health effects associated with biological, chemical and physical agents which may be present in food. For chemical agents, a dose-response assessment should be performed. For biological or physical agents, a dose-response assessment should be performed if the data are obtainable." In this stage, risk assessors should describe the nature and extent of the adverse health effects known to be associated with the specific hazard. Using toxicity studies and epidemiological data, a dose-response relationship should be established between different levels of exposure to the hazard and the likelihood of different adverse health effects.

Exposure Assessment

"The qualitative and/or quantitative evaluation of the likely intake of biological, chemical, and physical agents via food as well as exposures from other sources if relevant." This step characterizes the amount of hazard that is consumed by various members of the exposure populations. Taking into account the food consumption patterns of the target population and levels of hazard in all steps of the production process, an exposure assessment examines the exposure to the hazard over a particular period of time in foods that are actually consumed. The assessment should also account for varying levels of hazard throughout production to estimate the likely hazard level at point of consumption.

Risk Characterization

"The qualitative and quantitative estimation, including attendant uncertainties, of the probability of occurrence and severity of known or potential adverse health effects in a given population based on hazard identification, hazard characterization and exposure assessment." During this stage, estimates of risk are generated from the outputs of hazard identification, hazard characterization, and exposure assessment. A proper risk characterization should take into account multiple degrees of uncertainty and variability.

Risk Communication

Risk communication is defined for the purposes of the Codex Alimentarius Commission as "The interactive exchange of information and opinions throughout the risk analysis process concerning hazards and risks, risk-related factors and risk perceptions, among risk assessors, risk managers, consumers, industry, the academic community and other interested parties, including the explanation of risk assessment findings and the basis of risk management decisions."

Codex Alimentarius Commission

The Codex Alimentarius Commission was created in 1963 by the Food Agriculture Organization (FAO) and the World Health Organization (WHO) to develop food standards, guidelines and related texts such as codes of practice under the Joint FAO/WHO Food Standards Programme. The main purposes of this Programme are protecting health of the consumers and ensuring fair trade practices in the food trade, and promoting coordination of all food standards work undertaken by international governmental and non-governmental organizations.

HAZARD ANALYSIS AND CRITICAL CONTROL POINTS

Hazard analysis and critical control points, or HACCP, is a systematic preventive approach to food safety from biological, chemical, and physical hazards in production processes that can cause the finished product to be unsafe and designs measures to reduce these risks to a safe level. In this manner, HACCP attempts to avoid hazards rather than attempting to inspect finished products for the effects of those hazards. The HACCP system can be used at all stages of a food chain, from food production and preparation processes including packaging, distribution, etc. The Food and Drug Administration (FDA) and the United States Department of Agriculture (USDA) require mandatory HACCP programs for juice and meat as an effective approach to food safety and protecting public health. Meat HACCP systems are regulated by the USDA, while seafood and juice are regulated by the FDA. All other food companies in the United States that are required to register with the FDA under the Public Health Security and Bioterrorism Preparedness and Response Act of 2002, as well as firms outside the US that export food to the US, are transitioning to mandatory hazard analysis and risk-based preventive controls (HARPC) plans.

HACCP is believed to stem from a production process monitoring used during World War II because traditional "end of the pipe" testing on artillery shells' firing mechanisms could not be performed, and a large percentage of the artillery shells made at the time were either duds or misfiring. HACCP itself was conceived in the 1960s when the US National Aeronautics and Space Administration (NASA) asked Pillsbury to design and manufacture the first foods for space flights. Since then, HACCP has been recognized internationally as a logical tool for adapting traditional inspection methods to a modern, science-based, food safety system. Based on risk-assessment, HACCP plans allow both industry and government to allocate their resources efficiently in establishing and auditing safe food production practices. In 1994, the organization International HACCP Alliance was established, initially to assist the US meat and poultry industries with implementing HACCP, and now its membership has been spread over other professional and industrial areas.

Hence, HACCP has been increasingly applied to industries other than food, such as cosmetics and pharmaceuticals. This method, which in effect seeks to plan out unsafe practices based on science, differs from traditional "produce and sort" quality control methods that do nothing to prevent hazards from occurring and must identify them at the end of the process. HACCP is focused only on the health safety issues of a product and not the quality of the product, yet HACCP principles are the basis of most food quality and safety assurance systems. In the United States, HACCP compliance is regulated by 21 CFR part 120 and 123. Similarly, FAO and WHO published a guideline for all governments to handle the issue in small and less developed food businesses.

In the early 1960s, a collaborated effort between the Pillsbury Company, NASA, and the U.S. Army Laboratories began with the objective to provide safe food for space expeditions. People involved in this collaboration included Herbert Hollander, Mary Klicka, and Hamed El-Bisi of the United States Army Laboratories in Natick, Massachusetts, Paul A. Lachance of the Manned Spacecraft Center in Houston, Texas, and Howard E. Baumann representing Pillsbury as its lead scientist.

To ensure that the food sent to space was safe, Lachance imposed strict microbial requirements, including pathogen limits (including *E. coli*, *Salmonella*, and *Clostridium botulinum*). Using the traditional end product testing method, it was soon realized that almost all of the food manufactured was being used for testing and very little was left for actual use. Therefore, a new approach was needed.

NASA's own requirements for critical control points (CCP) in engineering management would be used as a guide for food safety. CCP derived from failure mode and effects analysis (FMEA) from NASA via the munitions industry to test weapon and engineering system reliability. Using that information, NASA and Pillsbury required contractors to identify "critical failure areas" and eliminate them from the system, a first in the food industry then. Baumann, a microbiologist by training, was so pleased with Pillsbury's experience in the space program that he advocated for his company to adopt what would become HACCP at Pillsbury.

Soon, Pillsbury was confronted with a food safety issue of its own when glass contamination was found in farina, a cereal commonly used in infant food. Baumann's leadership promoted HACCP in Pillsbury for producing commercial foods, and applied to its own food production. This led to a panel discussion at the 1971 National Conference on Food Protection that included examining CCPs and good manufacturing practices in producing safe foods. Several botulism cases were attributed to under-processed low-acid canned foods in 1970–71. The United States Food and Drug Administration (FDA) asked Pillsbury to organize and conduct a training program on the inspection of canned foods for FDA inspectors. This 21-day program was first held in September 1972 with 11 days of classroom lecture and 10 days of canning plant evaluations. Canned food regulations (21 CFR 108, 21 CFR 110, 21 CFR 113, and 21 CFR 114) were first published in 1969. Pillsbury's training program to the FDA in 1969, titled "Food Safety through the Hazard Analysis and Critical Control Point System", was the first time that HACCP was used.

HACCP was initially set on three principles, now shown as principles one, two, and four. Pillsbury quickly adopted two more principles, numbers three and five, to its own company in 1975. It was further supported by the National Academy of Sciences (NAS) that governmental inspections by the FDA go from reviewing plant records to compliance with its HACCP system. A second proposal by the NAS led to the development of the National Advisory Committee on Microbiological Criteria for Foods (NACMCF) in 1987. NACMCF was initially responsible for defining HACCP's systems and guidelines for its application and were coordinated with the Codex Committee for Food Hygiene, that led to reports starting in 1992 and further harmonization in 1997. By 1997, the seven HACCP principles listed below became the standard. A year earlier, the American Society for Quality offered their first certifications for HACCP Auditors. (First known as Certified Quality Auditor-HACCP, they were changed to Certified HACCP Auditor (CHA) in 2004.).

HACCP expanded in all realms of the food industry, going into meat, poultry, seafood, dairy, and has spread now from the farm to the fork.

Principles

- Conduct a hazard analysis:
  - Plan to determine the food safety hazards and identify the preventive measures the plan can apply to control these hazards. A food safety hazard is any biological, chemical, or physical property that may cause a food to be unsafe for human consumption.
- Identify critical control points:
  - A critical control point (CCP) is a point, step, or procedure in a food manufacturing process at which control can be applied and, as a result, a food safety hazard can be prevented, eliminated, or reduced to an acceptable level.

- Establish critical limits for each critical control point:
    - A critical limit is the maximum or minimum value to which a physical, biological, or chemical hazard must be controlled at a critical control point to prevent, eliminate, or reduce that hazard to an acceptable level.
- Establish critical control point monitoring requirements:
    - Monitoring activities are necessary to ensure that the process is under control at each critical control point. In the United States, the FSIS requires that each monitoring procedure and its frequency be listed in the HACCP plan.
- Establish corrective actions:
    - These are actions to be taken when monitoring indicates a deviation from an established critical limit. The final rule requires a plant's HACCP plan to identify the corrective actions to be taken if a critical limit is not met. Corrective actions are intended to ensure that no product is injurious to health or otherwise adulterated as a result if the deviation enters commerce.
- Establish procedures for ensuring the HACCP system is working as intended:
    - Validation ensures that the plants do what they were designed to do; that is, they are successful in ensuring the production of a safe product. Plants will be required to validate their own HACCP plans. FSIS will not approve HACCP plans in advance, but will review them for conformance with the final rule.
    - Verification ensures the HACCP plan is adequate, that is, working as intended. Verification procedures may include such activities as review of HACCP plans, CCP records, critical limits and microbial sampling and analysis. FSIS is requiring that the HACCP plan include verification tasks to be performed by plant personnel. Verification tasks would also be performed by FSIS inspectors. Both FSIS and industry will undertake microbial testing as one of several verification activities.
    - Verification also includes 'validation' – the process of finding evidence for the accuracy of the HACCP system (e.g. scientific evidence for critical limitations).
- Establish record keeping procedures:
    - The HACCP regulation requires that all plants maintain certain documents, including its hazard analysis and written HACCP plan, and records documenting the monitoring of critical control points, critical limits, verification activities, and the handling of processing deviations. Implementation involves monitoring, verifying, and validating of the daily work that is compliant with regulatory requirements in all stages all the time. The differences among those three types of work are given by Saskatchewan Agriculture and Food.

Application

- Fish and fishery products.
- Fresh-cut produce.
- Juice and nectar products.
- Food outlets.
- Meat and poultry products.
- School food and services.

Water Quality Management

The use of HACCP for water quality management was first proposed nearly 20 years ago. Thereafter, a number of water quality initiatives applied HACCP principles and steps to the control of infectious disease from water, and provided the basis for the Water Safety Plan (WSP) approach in the third edition of the World Health Organization (WHO) report, which has been described as "a way of adapting the HACCP approach to drinking water systems".

Water Quality Management Programme Guidelines

Programme Modernization: According to Ongley, 1998, a series of steps could be taken to execute a more useful transition – from technical programmes to policy to management decisions. Various aspects of the modernization process have been discussed by Ongley in ESCAP:

- Policy reform: A consultative process must define all the policy tenets and should review the execution of the said policy tenets.
- Legal reform: Legal reform with respect to water quality management is one of the most crucial elements. This could be addressed by the creation of national data standards as well as the creation of a national process to analyze and review collected data.
- Institutional reform: This is a complex issue and has got no simple answers. Still, there are some key principles that can be helpful for institutional reform in the light of water quality management. One of them is water quality monitoring as a service function. Apart from that, both technical efficiency and capacity issues emerge as major factors in a reformed water quality programmes.
- Technical reform: This is one area that garners the most attention as well as investment. However, such a kind of reform most targets facility modernization, including other co-factors like data programmes/networks, technical innovation, data management/data products and remediation.

HACCP for Building Water Systems

Hazards associated with water systems in buildings include physical, chemical and microbial hazards. In 2013, NSF International, a public health and safety NGO, established education, training and certification programs in HACCP for building water systems. The programs, developed with the guidance of subject matter experts Aaron Rosenblatt and William McCoy center on the use of HACCP principles adapted to the specific requirements of domestic (hot and cold) and utility (HVAC, etc.) water systems in buildings, to prevent plumbing-associated hazards from harming people. Hazards addressed include scalding, lead, and disinfection byproducts as well as a range of clinically important pathogens, such as *Legionella*, *Pseudomonas*, nontuberculous mycobacteria (NTM), *Acinetobacter*, *Elizabethkingia*, and *Naegleria*. Early adopters of HACCP for building water systems include leading healthcare institutions, notably Mayo Clinic in Rochester, MN.

ISO 22000

ISO 22000 is a standard designed to help augment HACCP on issues related to food safety. Although several companies, especially big ones, have either implemented or are on the point of implementing ISO 22000, there are many others which are hesitant to do so. The main reason behind that is the lack of information and the fear that the new standard is too demanding in terms of bureaucratic work.

FOOD TESTING STRIPS

Food testing strips are products that help determine whether or not food contains bacteria that can cause foodborne illness. These products can typically be used on food, water, and hard surfaces, and are often designed for quick and easy home and commercial use.

Categories

Currently, there are two categories of food testing strips on the market.

One type of food testing strip is an assay enzyme reactant test. This test requires the food testing strip to be dipped into a blended mixture of food or test samples, distilled water and a reagent. These strips are designed specifically to detect those strains of *E.coli* and *Salmonella* that are harmful to humans.

A second type of food testing strip is a gram-negative swab, which is usually administered directly to the food itself. Gram-negative swabs generally work faster than enzyme reactant strips, but they differ in that the gram-negative swabs are designed to detect a broad group of organisms, not just those that can cause foodborne illness in humans.

Usage

The enzyme reactant test strips react when the buffer solution breaks the bacterial wall. This breach releases enzymes, which react upon contact to the enzyme test strips.

The gram-negative reactant activates when components of the gram-negative cell wall or specific enzymes are present, causing the swab itself to change color. This is not directly indicative of the presence or absence of human pathogen in the test sample.

People are now working on new ways to enhance these pathogen strips with silk pills and new nano-fiber technology.

FOOD STORAGE

Yup'ik elevated food cache (*qulvarvik*), Hooper Bay, Alaska, 1929. Photograph by Edward S. Curtis.

U.S. Federal Emergency Management Agency (FEMA) food storage containers stacked on shipping pallets in Texas.

Food storage allows food to be eaten for some time (typically weeks to months) after harvest rather than solely immediately. It is both a traditional domestic skill and, in the form of food logistics, an important industrial and commercial activity. Food preservation, storage, and transport, including timely delivery to consumers, are important to food security, especially for the majority of people throughout the world who rely on others to produce their food. Food is stored by almost every human society and by many animals. Storing of food has several main purposes:

- Storage of harvested and processed plant and animal food products for distribution to consumers.
- Enabling a better balanced diet throughout the year.
- Reducing kitchen waste by preserving unused or uneaten food for later use.
- Preserving pantry food, such as spices or dry ingredients like rice and flour, for eventual use in cooking.

- Preparedness for catastrophes, emergencies and periods of food scarcity or famine.
- Religious reasons (Example: LDS Church leaders instruct church members to store food).
- Protection from animals or theft.

A new braided granary is inaugurated. Kapsiki, North Cameroon.

Domestic Food Storage

Tupperware kitchen storage containers designed for a variety of uses.

The safe storage of food for home use should strictly adhere to guidelines set out by reliable sources, such as the United States Department of Agriculture. These guidelines have been thoroughly researched by scientists to determine the best methods for reducing the real threat of food poisoning from unsafe food storage. It is also important to maintain proper kitchen hygiene, to reduce risks of bacteria or virus growth and food poisoning. The common food poisoning illnesses include Listeriosis, Mycotoxicosis, Salmonellosis, E. coli, Staphylococcal food poisoning and Botulism. There are many other organisms that can also cause food poisoning.

There are also safety guidelines available for the correct methods of home canning of food. For example, there are specific boiling times that apply depending upon whether pressure canning or waterbath canning is being used in the process. These safety guidelines are intended to reduce the growth of mold and bacteria and the threat of potentially-fatal food poisoning.

Plastic storage containers can be used to store food.

Food Storage Safety

Freezers and Thawing Food

Freezer temperature should be maintained below 0 °F (−18 °C). Food should never be thawed at room temperature, this increases the risk of bacterial and fungal growth and accordingly the risk of food poisoning. Once thawed, food should be used and never refrozen. Frozen food should be thawed using the following methods:

- Microwave oven,
- During cooking,
- In cold water (place food in watertight, plastic bag; change water every 30 minutes),
- In the refrigerator.

Throw out foods that have been warmer than 40 °F (4 °C) for more than 2 hours. If there is any doubt at all about the length of time the food has been defrosted at room temperature, it should be thrown out. Freezing does not destroy microbes present in food. Freezing at 0 °F does inactivate microbes (bacteria, yeasts and molds). However, once food has been thawed, these microbes can again become active. Microbes in thawed food can multiply to levels that can lead to foodborne illness. Thawed food should be handled according to the same guidelines as perishable fresh food.

Food frozen at 0 °F and below is preserved indefinitely. However, the quality of the food will deteriorate if it is frozen over a lengthy period. The United States Department of Agriculture, Food Safety and Inspection Service publishes a chart showing the suggested freezer storage time for common foods.

Refrigeration

It is important to note that safe food storage using refrigeration requires adhering to temperature guidelines:

For safety, it is important to verify the temperature of the refrigerator. Refrigerators should be set to maintain a temperature of 40 °F or below. Some refrigerators have built-in thermometers to measure their internal temperature. For those refrigerators without this feature, keep an appliance thermometer in the refrigerator to monitor the temperature. This can be critical in the event of a power outage. When the power goes back on, if the refrigerator is still 40 °F, the food is safe. Foods held at temperatures above 40 °F for more than 2 hours should not be consumed. Appliance thermometers are specifically designed to provide accuracy at cold temperatures. Be sure refrigerator/freezer doors are closed tightly at all times. Don't open refrigerator/freezer doors more often than necessary and close them as soon as possible.

Storage Times for Refrigerated Food

The United States Department of Agriculture, Food Safety and Inspection Service publishes recommended storage times for refrigerated food.

Storing Oils and Fats

Oils and fats can begin to go rancid quickly when not stored safely. Rancid cooking oils and fats do not often smell rancid until well after they have spoiled. Oxygen, light and heat all contribute to cooking oils becoming rancid. The higher the level of polyunsaturated fat that an oil contains, the faster it spoils. The percentage of polyunsaturated fat in some common cooking oils is: safflower (74%); sunflower (66%); corn (60%); soybean (37%); peanut (32%); canola (29%); olive (8%); coconut (5%).

To help preserve oils from rancidification, they should be stored in a dark place, stored in oxygen-safe, light-reducing containers (e.g. dark glass or metal). Once opened, oils should be refrigerated and used within a few weeks, when some types begin to go rancid. Unopened oils can have a storage life of up to one year, but some types have a shorter shelf-life even when unopened (such as sesame and flaxseed).

Dry Storage of Foods

Vegetables

The guidelines vary for safe storage of vegetables under dry conditions (without refrigerating or freezing). This is because different vegetables have different characteristics, for example, tomatoes contain a lot of water, while root vegetables such as carrots and potatoes contain less. These factors, and many others, affect the amount of time that a vegetable can be kept in dry storage, as well as the temperature needed to preserve its usefulness. The following guideline shows the required dry storage conditions:

- Cool and dry: Onion, List of Allium species.
- Cool and moist: Root vegetable, potato, cabbage.
- Warm and dry: Winter squash, pumpkin, sweet potatoes, dried hot peppers.

A large root cellar at the Oxon Hill Manor farm in Maryland.

Many cultures have developed innovative ways of preserving vegetables so that they can be stored for several months between harvest seasons. Techniques include pickling, home canning, food dehydration, or storage in a root cellar.

Grain

Grain, which includes dry kitchen ingredients such as flour, rice, millet, couscous, cornmeal, and so on, can be stored in rigid sealed containers to prevent moisture contamination or insect or rodent infestation. For kitchen use, glass containers are the most traditional method. During the 20th century plastic containers were introduced for kitchen use. They are now sold in a vast variety of sizes and designs.

Metal cans are used (in the United States the smallest practical grain storage uses closed-top #10 metal cans). Storage in grain sacks is ineffective; mold and pests destroy a 25 kg cloth sack of grain in a year, even if stored off the ground in a dry area. On the ground or damp concrete, grain can spoil in as little as three days, and the grain might have to be dried before it can be milled. Food stored under unsuitable conditions should not be purchased or used because of risk of spoilage. To test whether grain is still good, it can be sprouted. If it sprouts, it is still good, but if not, it should not be eaten. It may take up to a week for grains to sprout. When in doubt about the safety of the food, throw it out as quickly as possible.

Spices and Herbs

Spices and herbs are today often sold prepackaged in a way that is convenient for pantry storage. The packaging has dual purposes of both storing and dispensing the spices or herbs. They are sold in small glass or plastic containers or resealable plastic packaging. When spices or herbs are homegrown or bought in bulk, they can be stored at home in glass or plastic containers. They can be stored for extended periods, in some cases for years. However, after 6 months to a year, spices and herbs will gradually lose their flavour as oils they contain will slowly evaporate during storage.

Spices and herbs can be preserved in vinegar for short periods of up to a month without losing flavor, creating a flavoured vinegar.

Alternative methods for preserving herbs include freezing in water or unsalted butter. Herbs can be chopped and added to water in an ice cube tray. After freezing, the ice cubes are emptied into a plastic freezer bag for storing in the freezer. Herbs also can be stirred into a bowl with unsalted butter, then spread on wax paper and rolled into a cylinder shape. The wax paper roll containing the butter and herbs is then stored in a freezer, and can be cut off in the desired amount for cooking. Using either of these techniques, the herbs should be used within a year.

Meat

Unpreserved meat has only a relatively short life in storage. Perishable meats should be refrigerated, frozen, dried promptly or cured. Storage of fresh meats is a complex discipline that affects the costs, storage life and eating quality of the meat, and the appropriate techniques vary with the kind of meat and the particular requirements. For example, dry ageing techniques are sometimes used to tenderize gourmet meats by hanging them in carefully controlled environments for up to 21 days, while game animals of various kinds may be hung after shooting. Details depend on personal tastes and local traditions. Modern techniques of preparing meat for storage vary with the type of meat and special requirements of tenderness, flavour, hygiene, and economy.

Semi-dried meats like salamis and country style hams are processed first with salt, smoke, sugar, acid, or other "cures" then hung in cool dry storage for extended periods, sometimes exceeding a year. Some of the materials added during the curing of meats serve to reduce the risks of food poisoning from anaerobic bacteria such as species of Clostridium that release botulinum toxin that can cause botulism. Typical ingredients of curing agents that inhibit anaerobic bacteria include nitrates. Such salts are dangerously poisonous in their own right and must be added in carefully controlled quantities and according to proper techniques. Their proper use has however saved many lives and much food spoilage.

Like the semi-dried meats, most salted, smoked, and simply-dried meats of different kinds that once were staples in particular regions, now are largely luxury snacks or garnishes; examples include jerky, biltong, and varieties of pemmican, but ham and bacon for instance, still are staples in many communities.

Food Rotation

Food rotation is important to preserve freshness. When food is rotated, the food that has been in storage the longest is used first. As food is used, new food is added to the pantry to replace it; the essential rationale is to use the oldest food as soon as possible so that nothing is in storage too long and becomes unsafe to eat. Labelling food with paper labels on the storage container, marking the date that the container is placed in storage, can make this practice simpler. The best way to rotate food storage is to prepare meals with stored food on a daily basis.

For Emergency Preparation

Guides for surviving emergency conditions in many parts of the world recommend maintaining a store of essential foods; typically water, cereals, oil, dried milk, and protein rich foods such as beans, lentils, tinned meat and fish. A food storage calculator can be used to help determine how much of these staple foods a person would need to store in order to sustain life for one full year. In addition to storing the basic food items many people choose to supplement their food storage with frozen or preserved garden-grown fruits and vegetables and freeze-dried or canned produce. An unvarying diet of staple foods prepared in the same manner can cause appetite exhaustion, leading to less caloric intake. Another benefit to having a basic supply of food storage in the home is for the potential cost savings. Costs of dry bulk foods (before preparation) are often considerably less than convenience and fresh foods purchased at local markets or supermarkets. There is a significant market in convenience foods for campers, such as dehydrated food products.

Commercial Food Logistics

Silos connected to a grain elevator on a farm in Israel.

Grain and beans are stored in tall grain elevators, almost always at a rail head near the point of production. The grain is shipped to a final user in hopper cars. In the former Soviet Union, where harvest was poorly controlled, grain was often irradiated at the point of production to suppress mold and insects. In the U.S., threshing and drying is performed in the field, and transport is nearly sterile and in large containers that effectively suppresses pest access, which eliminates the need for irradiation. At any given time, the U.S. usually has about two weeks worth of stored grains for the population.

Fresh fruits and vegetables are sometimes packed in plastic packages and cups for fresh premium markets, or placed in large plastic tubs for sauce and soup processors. Fruits and vegetables are usually refrigerated at the earliest possible moment, and even so have a shelf life of two weeks or less.

In the United States, livestock is usually transported live, slaughtered at a major distribution point, hung and transported for two days to a week in refrigerated rail cars, and then butchered and sold locally. Before refrigerated rail cars, meat had to be transported live, and this placed its cost so high that only farmers and the wealthy could afford it every day. In Europe much meat is transported live and slaughtered close to the point of sale. In much of Africa and Asia most meat is for local populations is raised, slaughtered and eaten locally, which is believed to be less stressful for the animals involved and minimizes meat storage needs. In Australia and New Zealand, where a large proportion of meat production is for export, meat enters the cold chain early, being stored in large freezer plants before being shipped overseas in freezer ships.

SAFE FOOD HANDLING

When certain disease-causing bacteria or pathogens contaminate food, they can cause foodborne illness, often called "food poisoning." The Federal government estimates that there are about 48 million cases of foodborne illness annually – the equivalent of sickening 1 in 6 Americans each year. And each year, these illnesses result in an estimated 128,000 hospitalizations and 3,000 deaths.

Know the Symptoms

Consuming dangerous foodborne bacteria will usually cause illness within 1 to 3 days of eating the contaminated food. However, sickness can also occur within 20 minutes or up to 6 weeks later. Symptoms of foodborne illness can include: vomiting, diarrhea, and abdominal pain – and flu-like symptoms, such as fever, headache, and body ache.

Handle Foods Safely

Although most healthy people will recover from a foodborne illness within a short period of time, some can develop chronic, severe, or even life-threatening health problems. In addition, some people are at a higher risk for developing foodborne illness, including pregnant women, young children, older adults, and people with weakened immune systems (such as transplant patients and individuals with HIV/AIDS, cancer, or diabetes). To keep your family safer from food poisoning, follow these four simple steps: clean, separate, cook, and chill.

Clean

Wash Hands and Surfaces

- Wash your hands with warm water and soap for at least 20 seconds before and after handling food and after using the bathroom, changing diapers, and handling pets.
- Wash your cutting boards, dishes, utensils, and counter tops with hot soapy water after preparing each food item.
- Consider using paper towels to clean up kitchen surfaces. If you use cloth towels, launder them often in the hot cycle.
- Rinse fresh fruits and vegetables under running tap water, including those with skins and rinds that are not eaten. Scrub firm produce with a clean produce brush.
- With canned goods, remember to clean lids before opening.

Separate

Separate Raw Meats from other Foods

- Separate raw meat, poultry, seafood, and eggs from other foods in your grocery shopping cart, grocery bags, and refrigerator.
- Use one cutting board for fresh produce and a separate one for raw meat, poultry, and seafood.
- Never place cooked food on a plate that previously held raw meat, poultry, seafood, or eggs unless the plate has been washed in hot, soapy water.
- Don't reuse marinades used on raw foods unless you bring them to a boil first.

Cook

Cook to the Right Temperature

- Color and texture are unreliable indicators of safety. Using a food thermometer is the only way to ensure the safety of meat, poultry, seafood, and egg products for all cooking methods. These foods must be cooked to a safe minimum internal temperature to destroy any harmful bacteria.
- Cook eggs until the yolk and white are firm. Only use recipes in which eggs are cooked or heated thoroughly.
- When cooking in a microwave oven, cover food, stir, and rotate for even cooking. If there is no turntable, rotate the dish by hand once or twice during cooking. Always allow standing time, which completes the cooking, before checking the internal temperature with a food thermometer.

- Bring sauces, soups and gravy to a boil when reheating.

Chill

Refrigerate Foods Promptly

- Use an appliance thermometer to be sure the temperature is consistently 40 °F or below and the freezer temperature is 0 °F or below.
- Refrigerate or freeze meat, poultry, eggs, seafood, and other perishables within 2 hours of cooking or purchasing. Refrigerate within 1 hour if the temperature outside is above 90 °F.
- Never thaw food at room temperature, such as on the counter top. There are three safe ways to defrost food: in the refrigerator, in cold water, and in the microwave. Food thawed in cold water or in the microwave should be cooked immediately.
- Always marinate food in the refrigerator.
- Divide large amounts of leftovers into shallow containers for quicker cooling in the refrigerator.

FROZEN FOODS

Freezing food preserves it from the time it is prepared to the time it is eaten. Since early times, farmers, fishermen, and trappers have preserved grains and produce in unheated buildings during the winter season. Freezing food slows down decomposition by turning residual moisture into ice, inhibiting the growth of most bacterial species. In the food commodity industry, there are two processes: mechanical and cryogenic (or flash freezing). The freezing kinetics is important to preserve the food quality and texture. Quicker freezing generates smaller ice crystals and maintains cellular structure. Cryogenic freezing is the quickest freezing technology available due to the ultra low liquid nitrogen temperature −196 °C (−320 °F).

A frozen processed foods aisle at a supermarket in Canada.

Preserving food in domestic kitchens during modern times is achieved using household freezers. Accepted advice to householders was to freeze food on the day of purchase. An initiative by a supermarket group in 2012 (backed by the UK's Waste & Resources Action Programme) promotes the freezing of food "as soon as possible up to the product's 'use by' date". The Food Standards Agency was reported as supporting the change, provided the food had been stored correctly up to that time.

Cutting frozen tuna using a bandsaw in the Tsukiji fish market in Tokyo, Japan.

Preservatives

Frozen products do not require any added preservatives because microorganisms do not grow when the temperature of the food is below −9.5 °C (15 °F), which is sufficient on its own in preventing food spoilage. Long-term preservation of food may call for food storage at even lower temperatures. Carboxymethylcellulose (CMC), a tasteless and odorless stabilizer, is typically added to frozen food because it does not adulterate the quality of the product.

Technology

The freezing technique itself, just like the frozen food market, is developing to become faster, more efficient and more cost-effective.

Mechanical freezers were the first to be used in the food industry and are used in the vast majority of freezing/refrigerating lines. They function by circulating a refrigerant, normally ammonia, around the system, which withdraws heat from the food product. This heat is then transferred to a condenser and dissipated into air or water. The refrigerant itself, now a high pressure, hot liquid, is directed into an evaporator.

As it passes through an expansion valve, it is cooled and then vaporises into a gaseous state. Now a low pressure, low temperature gas again, it can be reintroduced into the system.

Cryogenic (or flash freezing) of food is a more recent development, but is used by many leading food manufacturers all over the world. Cryogenic equipment uses very low temperature gases – usually liquid nitrogen or solid carbon dioxide – which are applied directly to the food product.

Packaging

Frozen food packaging must maintain its integrity throughout filling, sealing, freezing, storage, transportation, thawing, and often cooking. As many frozen foods are cooked in a microwave oven, manufacturers have developed packaging that can go straight from freezer to the microwave.

In 1974, the first differential heating container (DHC) was sold to the public. A DHC is a sleeve of metal designed to allow frozen foods to receive the correct amount of heat. Various sized apertures were positioned around the sleeve. The consumer would put the frozen dinner into the sleeve according to what needed the most heat. This ensured proper cooking.

Today there are multiple options for packaging frozen foods. Boxes, cartons, bags, pouches, Boil-in-Bags, lidded trays and pans, crystallized PET trays, and composite and plastic cans.

Scientists are continually researching new aspects of frozen food packaging. Active packaging offers a host of new technologies that can actively sense and then neutralize the presence of bacteria or other harmful species. Active packaging can extend shelf-life, maintain product safety, and help preserve the food over a longer period of time. Several functions of active packaging are being researched:

- Oxygen scavengers,
- Time Temperature Indicators and digital temperature data loggers,
- Antimicrobials,
- Carbon Dioxide controllers,
- Microwave susceptors,
- Moisture control: Water activity, Moisture vapor transmission rate, etc.,
- Flavor enhancers,
- Odor generators,

- Oxygen-permeable films,
- Oxygen generators.

Effects on Nutrients

Vitamin Content of Frozen Foods

- Vitamin C: Usually lost in a higher concentration than any other vitamin. A study was performed on peas to determine the cause of vitamin C loss. A vitamin loss of ten percent occurred during the blanching phase with the rest of the loss occurring during the cooling and washing stages. The vitamin loss was not actually accredited to the freezing process. Another experiment was performed involving peas and lima beans. Frozen and canned vegetables were both used in the experiment. The frozen vegetables were stored at −23 °C (−10 °F) and the canned vegetables were stored at room temperature 24 °C (75 °F). After 0, 3, 6, and 12 months of storage, the vegetables were analyzed with and without cooking. O'Hara, the scientist performing the experiment said, "From the view point of the vitamin content of the two vegetables when they were ready for the plate of the consumer, there did not appear to be any marked advantages attributable to method of preservation, frozen storage, processed in a tin, or processed in glass."
- Vitamin B_1 (Thiamin): A vitamin loss of 25 percent is normal. Thiamin is easily soluble in water and is destroyed by heat.
- Vitamin B_2 (Riboflavin): Not much research has been done to see how much freezing affects Riboflavin levels. Studies that have been performed are inconclusive; one study found an 18 percent vitamin loss in green vegetables, while another determined a 4 percent loss. It is commonly accepted that the loss of Riboflavin has to do with the preparation for freezing rather than the actual freezing process itself.
- Vitamin A (Carotene): There is little loss of carotene during preparation for freezing and freezing of most vegetables. Much of the vitamin loss is incurred during the extended storage period.

Effectiveness

Freezing is an effective form of food preservation because the pathogens that cause food spoilage are killed or do not grow very rapidly at reduced temperatures. The process is less effective in food preservation than are thermal techniques, such as boiling, because pathogens are more likely to be able to survive cold temperatures rather than hot temperatures. One of the problems surrounding the use of freezing as a method of food preservation is the danger that pathogens deactivated (but not killed) by the process will once again become active when the frozen food thaws.

Foods may be preserved for several months by freezing. Long-term frozen storage requires a constant temperature of −18 °C (0 °F) or less.

A frozen food warehouse at McMurdo Station, Antarctica.

Defrosting

To be used, many cooked foods that have been previously frozen require defrosting prior to consumption. Preferably, some frozen meats should be defrosted prior to cooking to achieve the best outcome: cooked through evenly and of good texture.

Ideally, most frozen foods should be defrosted in a refrigerator to avoid significant growth of pathogens. However, this can take considerable time.

Food is often defrosted in one of several ways:

- At room temperature; this is dangerous since the outside may be defrosted while the inside remains frozen.
- In a refrigerator.
- In a microwave oven.
- Wrapped in plastic and placed in cold water or under cold running water.

People sometimes defrost frozen foods at room temperature because of time constraints or ignorance; such foods should be promptly consumed after cooking or discarded and never be refrozen or refrigerated since pathogens are not killed by the freezing process.

Quality

The speed of the freezing has a direct impact on the size and the number of ice crystals formed within a food product's cells and extracellular space. Slow freezing leads to fewer but larger ice crystals while fast freezing leads to smaller but more numerous ice crystals. Large ice crystals can puncture the walls of the cells of the food product which will cause a degradation of the texture of the product as well as the loss of its natural juices during thawing. That is why there will be a qualitative difference observed between food products frozen by ventilated mechanical freezing, non-ventilated mechanical freezing or cryogenic freezing with liquid nitrogen.

FOOD PACKAGING

Food packaging is packaging for food. A package provides protection, tampering resistance, and special physical, chemical, or biological needs. It may bear a nutrition facts label and other information about food being offered for sale.

Testing modified atmosphere in a plastic bag of carrots.

Functions

Packaging and package labeling have several objectives:

- Physical protection: The food enclosed in the package may require protection from, among, shock, vibration, compression, temperature, bacteria, etc.
- Barrier protection: A barrier from oxygen, water vapor, dust, etc., is often required. Permeation is a critical factor in design. Some packages contain desiccants or oxygen absorbers to help extend shelf life. Modified atmospheres or controlled atmospheres are also maintained in some food packages. Keeping the contents clean, fresh, and safe for the intended shelf life is a primary function.
- Containment or agglomeration: Small items are typically grouped together in one package to allow efficient handling. Liquids, powders, and granular materials need containment.

- Information transmission: Packages and labels communicate how to use, transport, recycle, or dispose of the package or product. Some types of information are required by governments.
- Marketing: The packaging and labels can be used by marketers to encourage potential buyers to purchase the product. Aesthetically pleasing and eye-appealing food presentations can encourage people to consider the contents. Package design has been an important and constantly evolving phenomenon for several decades. Marketing communications and graphic design are applied to the surface of the package and (in many cases) the point of sale display. The colour of the package plays a significant role in evoking emotions that persuade the consumer to make the purchase.
- Security: Packaging can play an important role in reducing the security risks of shipment. Packages can be made with improved tamper resistance to deter tampering and also can have tamper-evident features to help indicate tampering. Packages can be engineered to help reduce the risks of package pilferage; some package constructions are more resistant to pilferage and some have pilfer-indicating seals. Packages may include authentication seals to help indicate that the package and contents are not counterfeit. Packages also can include anti-theft devices, such as dye packs, RFID tags, or electronic article surveillance tags, that can be activated or detected by devices at exit points and require specialized tools to deactivate. Using packaging in this way is a means of retail loss prevention.
- Convenience: Packages can have features which add convenience in distribution, handling, stacking, display, sale, opening, reclosing, use, and reuse.
- Portion control: Single-serving packaging has a precise amount of contents to control usage. Bulk commodities (such as salt) can be divided into packages that are a more suitable size for individual households. It also aids the control of inventory: selling sealed one-liter bottles of milk, rather than having people bring their own bottles to fill themselves.

Types

The previous materials are fashioned into different types of food packages and containers such as:

Packaging Type		
Aseptic processing	Primary	Liquid whole eggs or dairy products.
Trays	Primary	Portion of fish or meat.
Bags	Primary	Potato chips, apples, rice.
Boxes	Secondary	Corrugated box of primary packages: box of cereal cartons, frozen pizzas.
Cans	Primary	Can of tomato soup.

Cartons, coated paper	Primary	Carton of eggs, milk or juice cartons.
Flexible packaging	Primary	Bagged salad.
Pallets	Tertiary	A series of boxes on a single pallet used to transport from the manufacturing plant to a distribution center.
Wrappers	Tertiary	Used to wrap the boxes on the pallet for transport.

Primary packaging is the main package that holds the food that is being processed. Secondary packaging combines the primary packages into one box being made. Tertiary packaging combines all of the secondary packages into one pallet.

A choice of packaging machinery requires consideration of technical capabilities, labor requirements, worker safety, maintainability, serviceability, reliability, ability to integrate into the packaging line, capital cost, floorspace, flexibility (change-over, materials, etc.), energy usage, quality of outgoing packages, qualifications (for food, pharmaceuticals, etc.), throughput, efficiency, productivity, and ergonomics, at a minimum.

Packaging machines may be of the following general types:

- Autocoding label and date verification.
- Blister-, skin- and vacuum-packaging machines.
- Capping, over-capping, lidding, closing, seaming and sealing machines.
- Cartoning machines.
- Case and tray forming, packing, unpacking, closing and sealing Machines.
- Check weighing machines.
- Cleaning, sterilizing, cooling and drying machines.
- Conveying and accumulating machines.
- Feeding, orienting, and placing machines.
- Filling machines for liquid and powdered products.
- Package filling and closing Machines.
- Form, fill and seal machines.
- Inspecting, detecting and checkweighing machines.
- Palletizing, depalletizing, and pallet unitizing machines.
- Labeling, marking, and other product identification machines.
- Wrapping machines.
- Converting machines.

Reducing Food Packaging

Reduced packaging and sustainable packaging are becoming more frequent. The motivations can be government regulations, consumer pressure, retailer pressure, and cost control. Reduced packaging often saves packaging costs.

In the UK, a Local Government Association survey produced by the British Market Research Bureau compared a range of outlets to buy 29 common food items and found that small local retailers and market traders "produced less packaging and more that could be recycled than the larger supermarkets".

Recycling of Food Packaging

After use, organic matter that is still in the food packaging needs to be separated from the packaging. This may also require rinsing of the food packaging.

Food packaging is created through the use of a wide variety of plastics and metals, papers, and glass materials. Recycling these products differs from the act of literally reusing them in the manner that the recycling process has its own algorithm which includes collecting, sourcing, processing, manufacturing and marketing these products. According to the Environmental Protection Agency of the United States the recycling rate has been steadily on the rise with data reporting that in 2005 40% of the food packaging and containers that were created were recycled and not just thrown away.

Trends in Food Packaging

- Numerous reports industry associations agree that use of smart indicators will increase. There are a number of different indicators with different benefits for food producers, consumers and retailers.
- Temperature recorders are used to monitor products shipped in a cold chain and to help validate the cold chain. Digital temperature data loggers measure and record the temperature history of food shipments. They sometimes have temperatures displayed on the indicator or have other output (lights, etc.): The data from a shipment can be downloaded (cable, RFID, etc.) to a computer for further analysis. These help identify if there has been temperature abuse of products and can help determine the remaining shelf life. They can also help determine the time of temperature extremes during shipment so corrective measures can be taken.
    - Time temperature indicators integrate the time and temperature experienced by the indicator and adjacent foods. Some use chemical reactions that result in a color change while others use the migration of a dye through a filter media. To the degree that these physical changes in the indicator match the degradation rate of the food, the indicator can help indicate probable food degradation.

- Radio Frequency Identification is applied to food packages for supply chain control and has shown a significant benefit in allowing food producers and retailers create full real time visibility of their supply chain.
- Plastic packaging being used is usually non-biodegradable due to possible interactions with the food. Also, biodegradable polymers often require special composting conditions to properly degrade. Normal sealed landfill conditions do not promote biodegradation. Biodegradable plastics includes biodegradable films and coatings synthesized from organic materials and microbial polymers. Some package materials are edible. For example, pharmaceuticals are sometimes in capsules made of gelatin, starch, potato or other materials. Newer bioplastics, films and products are being developed.
- Barcodes have been used for decades in packaging many products. 2D barcodes used in Autocoding are increasingly applied to food packaging to ensure products are correctly packaged and date coded.
- The ability of a package to fully empty or dispense a viscous food is somewhat dependent on the surface energy of the inner walls of the container. The use of superhydrophobic surfaces is useful but can be further improved by using new lubricant-impregnated surfaces.

Food Safety and Public Health

It is critical to maintain food safety during processing, packaging, storage, logistics (including cold chain), sale, and use. Conformance to applicable regulations is mandatory. Some are country specific such as the US Food and Drug Administration and the US Department of Agriculture; others are regional such as the European Food Safety Authority. Certification programs such as the Global Food Safety Initiative are sometimes used. Food packaging considerations may include: use of hazard analysis and critical control points, verification and validation protocols, Good manufacturing practices, use of an effective quality management system, track and trace systems, and requirements for label content. Special food contact materials are used when the package is in direct contact with the food product. Depending on the packaging operation and the food, packaging machinery often needs specified daily wash-down and cleaning procedures.

Health risks of materials and chemicals used in food packaging need to be carefully controlled. Carcinogens, toxic chemicals, mutagens etc. need to be eliminated from food contact and potential migration into foods.

Manufacturing

Packaging lines may have a variety of equipment types: integration of automated systems can be a challenge. All aspects of food production, including packaging, are tightly controlled and have regulatory requirements. Uniformity, cleanliness and other requirements are needed to maintain Good Manufacturing Practices.

Product safety management is vital. A complete Quality Management System must be in place. Hazard analysis and critical control points is one methodology which has been proven useful. Verification and validation involves collecting documentary evidence of all aspects of compliance. Quality assurance extends beyond the packaging operations through distribution.

CUTTING FOOD WASTE AND MAINTAIN FOOD SAFETY

Food safety is a major concern. The Centers for Disease Control and Prevention (CDC) estimates that there are about 48 million cases of foodborne illness annually – the equivalent of sickening 1 in 6 Americans each year. And each year these illnesses result in an estimated 128,000 hospitalizations and 3,000 deaths.

Food waste is also a major concern. Wasted food is a huge challenge to our natural resources, our environment, and our pocketbooks.

The environment? Organic waste, mostly food, is the second biggest component of landfills, and landfills are the third largest source of methane emissions. Methane is a major factor in global warming because it is so effective at absorbing the sun's heat, which warms the atmosphere.

And, finally, our pocketbooks: Between 30 and 40 percent of food in the United States goes uneaten – as much as 20 pounds of food per person per month. That means Americans are throwing out the equivalent of $165 billion in food each year.

How Food Waste and Food Safety are Connected

The major sources of food waste in the United States are the food industry and consumers. Within the food industry, waste occurs at every step — on the farm and with packers, processors, distributors, and retailers. Some of it is the result of economic forces, some of management problems, and some is caused simply by dumping products that are less than perfect in appearance.

Food Product Dating and Food Waste

Food waste by consumers may result from a misunderstanding of what the phrases on product date labels mean, along with uncertainty about storage of perishable foods. Confusion over date labeling accounts for an estimated 20 percent of consumer food waste.

What are Food Product Dates?

Many consumers misunderstand the purpose and meaning of the date labels that often appear on packaged foods. Confusion over date labeling accounts for an estimated 20 percent of consumer food waste.

Except for infant formula, manufacturers are not required by Federal law or regulation to place quality-based date labels on packaged food.

There are no uniform or universally accepted descriptions used on food labels for open dating (calendar dates) in the United States. As a result, there are a wide variety of phrases used for product dating.

FDA supports efforts by the food industry to make "Best if Used By" the standard phrase to indicate the date when a product will be at its best flavor and quality. Consumers should examine foods for signs of spoilage that are past their "Best if used by" date. If the products have changed noticeably in color, consistency or texture, consumers may want to avoid eating them. If you have questions or concerns about the quality, safety and labeling of the packaged foods you buy, you are encouraged to reach out to the company that produced the product. Many packaged foods provide the company's contact information on the package.

Manufacturers apply date labels at their own discretion and for a variety of reasons. The most common is to inform consumers and retailers of the date to which they can expect the food to retain its desired quality and flavor.

Industry is moving toward more uniform practices for date labeling of packaged foods. But, for now, consumers may see different phrases used for product dating, such as Sell By, Best By, Expires on, etc.

How to best to Store Perishables and How Long they will keep Safely?

The FoodKeeper, developed cooperatively by the U.S. Department of Agriculture, Cornell University and the Food Marketing Institute, is a complete guide to how long virtually every food available in the United States will keep in the pantry, in the refrigerator, and in the freezer. The Fresh Fruits section, for example, covers apples (3 weeks in the pantry, 4 – 6 weeks in the fridge, and — only if cooked — 8 months in the freezer) to pomegranates (2 – 5 days pantry, 1 – 3 months fridge, and 10 – 12 months freezer). The Meat, Poultry and Seafood sections are equally complete, and include smoked as well fresh products.

More ways to avoid wasting food:

- Be aware of how much food you throw away.
- Don't buy more food than can be used before it spoils.

- Plan meals and use shopping lists: Think about what you are buying and when it will be eaten. Check the fridge and pantry to avoid buying what you already have.
- Avoid impulse and bulk purchases, especially produce and dairy that have a limited shelf life. Promotions encouraging purchases of unusual or bulk products often result in consumers buying foods outside their typical needs or family preferences, and portions — potentially large portions — of these foods may end up in the trash.
- When eating out, become a more mindful eater: If you're not terribly hungry request smaller portions. Bring your leftovers home and refrigerate or freeze them within two hours, and check the Food Keeper to see how long they'll be safe to eat.
- Check the temperature setting of your fridge: Use a refrigerator thermometer to be sure the temperature is at 40° F or below to keep foods safe. The temperature of your freezer should be 0° F or below.
- Avoid "overpacking:" Cold air must circulate around refrigerated foods to keep them properly chilled.
- Wipe up spills immediately: It not only reduces the growth of Listeria bacteria (which can grow at refrigerator temperatures), cleaning up spills — especially drips from thawing meats — will help prevent "cross-contamination," where bacteria from one food spread to another.
- Keep it covered: Store refrigerated foods in covered containers or sealed storage bags, and check leftovers daily for spoilage.
- Refrigerate peeled or cut veggies for freshness and to keep them from going bad.
- Use your freezer! Freezing is a great way to store most foods to keep them from going bad until you are ready to eat them. The FoodKeeper has information on how long most common foods can be stored in the freezer.
- Check your fridge often to keep track of what you have and what needs to be used. Eat or freeze items before you need to throw them away.
- To keep foods safe when entertaining, remember the 2-Hour Rule: don't leave perishable foods out at room temperature for more than two hours, unless you're keeping hot foods hot and cold foods cold. If you're eating outdoors and the temperature is above 90° F, perishable foods shouldn't be left out for more than one hour.

References

- Food-safety-problems-and-solutions, significance-prevention-and-control-of-food-related-diseases: intechopen.com, Retrieved 28 April, 2020

- Safe-food-handling, food-buy-store-serve-safe-food: fda.gov, Retrieved 16 May, 2020
- Robertson, G. L. (2013). "Food Packaging: Principles & Practice". CRC Press. ISBN 978-1-4398-6241-4
- Hans-Jürgen Bässler und Frank Lehmann : Containment Technology: Progress in the Pharmaceutical and Food Processing Industry. Springer, Berlin 2013, ISBN 978-3642392917
- How-cut-food-waste-and-maintain-food-safety, food-consumers: fda.gov, Retrieved 19 August, 2020

5

Foodborne Illnesses

The consumption of contaminated food can cause many foodborne illnesses. Some of them are diarrhea, gastroenteritis, traveler's diarrhea, vibrio vulnificus, food poisoning, etc. This chapter closely examines these different foodborne illnesses to provide an extensive understanding of the subject.

FOOD POISONING

Foodborne illness, more commonly referred to as food poisoning, is the result of eating contaminated, spoiled, or toxic food. The most common symptoms of food poisoning include nausea, vomiting, and diarrhea.

Although it's quite uncomfortable, food poisoning isn't unusual.

Food Poisoning Symptoms

If you have food poisoning, chances are it won't go undetected. Symptoms can vary depending on the source of the infection. The length of time it takes for symptoms to appear also depends on the source of the infection, but it can range from as little as 1 hour to as long as 28 days. Common cases of food poisoning will typically include at least three of the following symptoms:

- Abdominal cramps,
- Diarrhea,
- Vomiting,
- Loss of appetite,
- Mild fever,
- Weakness,
- Nausea,
- Headaches.

Symptoms of potentially life-threatening food poisoning include:

- Diarrhea persisting for more than three days.
- A fever higher than 101.5 °F.
- Difficulty seeing or speaking.
- Symptoms of severe dehydration, which may include dry mouth, passing little to no urine, and difficulty keeping fluids down.
- Bloody urine.

If you experience any of these symptoms, you should contact your doctor immediately.

What Causes Food Poisoning?

Most food poisoning can be traced to one of the following three major causes:

Bacteria

Bacteria is by far the most prevalent cause of food poisoning. When thinking of dangerous bacteria, names like E. coli, Listeria, and Salmonellacome to mind for good reason. Salmonella is by far the biggest culprit of serious food poisoning cases in the United States. According to the CDCTrusted Source, an estimated 1,000,000 cases of food poisoning, including nearly 20,000 hospitalizations, can be traced to salmonella infection annually. Campylobacter and C. botulinum (botulism)are two lesser-known and potentially lethal bacteria that can lurk in our food.

Parasites

Food poisoning caused by parasites is not as common as food poisoning caused by bacteria, but parasites spread through food are still very dangerous. Toxoplasmais the parasite seen most often in cases of food poisoning. It's typically found in cat litter boxes. Parasites can live in your digestive tract undetected for years. However, people with weakened immune systems and pregnant women risk serious side effects if parasites take up residence in their intestines.

Viruses

Food poisoning can also be caused by a virus. The norovirus, also known as the Norwalk virus, causes over 19 million casesTrusted Source of food poisoning each year. In rare cases, it can be fatal. Sapovirus, rotavirus, and astrovirus bring on similar symptoms, but they're less common. Hepatitis A virus is a serious condition that can be transmitted through food.

How does Food become Contaminated?

Pathogens can be found on almost all of the food that humans eat. However, heat from cooking usually kills pathogens on food before it reaches our plate. Foods eaten raw are common sources of food poisoning because they don't go through the cooking process.

Occasionally, food will come in contact with the organisms in fecal matter. This most commonly happens when a person preparing food doesn't wash their hands before cooking.

Meat, eggs, and dairy products are frequently contaminated. Water may also be contaminated with organisms that cause illness.

Who is at Risk for Food Poisoning?

Anyone can come down with food poisoning. Statistically speaking, nearly everyone will come down with food poisoning at least once in their lives.

There are some populations that are more at risk than others. Anyone with a suppressed immune system or an auto-immune disease may have a greater risk of infection and a greater risk of complications resulting from food poisoning.

According to the Mayo Clinic, pregnant women are more at risk because their bodies are coping with changes to their metabolism and circulatory system during pregnancy. Elderly individuals also face a greater risk of contracting food poisoning because their immune systems may not respond quickly to infectious organisms. Children are also considered an at-risk population because their immune systems aren't as developed as those of adults. Young children are more easily affected by dehydration from vomiting and diarrhea.

How is Food Poisoning Diagnosed?

Your doctor may be able to diagnose the type of food poisoning based on your symptoms. In severe cases, blood tests, stool tests, and tests on food that you have eaten may be conducted to determine what is responsible for the food poisoning. Your doctor may also use a urine test to evaluate whether an individual is dehydrated as a result of food poisoning.

How can Food Poisoning be Prevented?

The best way to prevent food poisoning is to handle your food safely and to avoid any food that may be unsafe.

Some foods are more likely to cause food poisoning because of the way they're produced and prepared. Meat, poultry, eggs, and shellfish may harbor infectious agents that are killed during cooking. If these foods are eaten in their raw form, not cooked properly, or if hands and surfaces are not cleaned after contact, food poisoning can occur.

Other foods that are likely to cause food poisoning include:

- Sushi and other fish products that are served raw or undercooked.
- Deli meats and hot dogs that are not heated or cooked.
- Ground beef, which may contain meat from several animals.
- Unpasteurized milk, cheese, and juice.
- Raw, unwashed fruits and vegetables.

Always wash your hands before cooking or eating food. Make sure that your food is properly sealed and stored. Thoroughly cook meat and eggs. Anything that comes in contact with raw products should be sanitized before using it to prepare other foods. Make sure to always wash fruits and vegetables before serving.

Food Poisoning caused by Bacteria

In the environment, there are polluting substances that can cause adverse reactions in human beings when entering the body through different ways (ingestion, inhalation, injection, or absorption). The main pollutants can be poisons, chemical compounds, toxic gases, and bacterial toxins. These can be found in different places and their effects depend on the dose and exposure time. Furthermore, foodborne diseases (FBDs) can cause disability; these diseases can be caused by toxins produced by bacteria or other toxic substances in the food, which can cause severe diarrhea, toxic shock syndrome, debilitating infections such as meningitis and even death. FBDs are transmitted through food contaminated with pathogenic microorganisms that have multiple factors of virulence, which gives them the ability to cause an infection; some bacterial genres can produce toxins directly in the food, but other genres can produce them once they have colonized the intestine. Among the pathogens involved in FBDs that are also considered to be toxigenic are Salmonella spp., Vibrio parahaemolyticus, Vibrio cholerae, Staphylococcus aureus, Clostridium botulinum, Clostridium perfringens, Bacillus cereus, Listeria monocytogenes. Foodborne diseases can be prevented and acute diarrhea syndromes, fever and even death from dehydration can be avoided, especially in children under the age of 5 and in immunocompromised people.

The main pollutants can be poisons, chemical compounds, toxic gases, and bacterial toxins. There are several diseases that human beings can acquire by ingesting some type of pollutants, for example, chemical contamination can lead to acute poisoning or long-term diseases such as cancer. Furthermore, foodborne diseases (FBDs) can cause disability; these diseases can be caused by the toxins produced by the bacteria or other toxic substances in food.

It is important to know that poisoning is the cause of morbidity and mortality worldwide. There are different types of intoxication: (a) intoxication caused by chemical substances (such as drugs, pesticides, heavy metals, gases, and solvents) where the patient has direct contact with the toxic substance, and (b) food poisoning, of which the transmission vehicle is contaminated food with pathogens or chemical products. Nowadays, chemical poisoning is a health problem; about six million chemicals are known, of which 80,000 to 100,000 are commonly used in different daily products. In 2006, thc World Health Organization (WHO) estimated that more than 25% of poisonings and 5% of cases of cancer, neuropsychiatric disorders, and vascular diseases worldwide were caused by chemical exposure.

It is difficult to diagnose chemical poisoning, since a chronological record of the patient's life is required, considering the exposure routes, dose, and time of exposure to the chemical. However, there are protocols that facilitate the diagnosis of chemical poisoning and how to treat incidents from chemical poisoning.

Furthermore, food poisoning or foodborne disease (FBD) is one of the main problems in public health worldwide. According to the WHO, each year 600 million people around the world, or 1 out of 10, become ill after consuming contaminated food. Among all these people, 420,000 die, including 125,000 children under 5 years of age, due to the vulnerability of this population to develop a diarrheal syndrome, about 43% of FBDs occur in these patients. About 70% of FBDs result from food contaminated with a microorganism.

Among the microorganisms causing FBDs are bacteria that have different virulence factors that give them the ability to cause a disease; among these factors, we can find toxins that can be produced in food or once the pathogen has colonized the digestive tract.

Types of Bacterial Toxins

A bacterial toxin is a macromolecule mainly of protein origin, which can cause toxic damage in a specific organ of the host. Toxins can be divided in endotoxins and exotoxins:

- Endotoxins or lipopolysaccharides (LPS): These are the components of the outer membrane of the Gram-negative bacteria; they are considered the most important antigen of the bacteria; they are released into the medium after different processes such as lysis and cell division. This endotoxin is capable of causing endotoxic shock and tissue damage.
- LPS are formed by three regions:
    - Lipid A is a glycolipid formed by a disaccharide (glucosamine) bound to fatty acids, that are usually capric, lauric, myristic, palmitic, and stearic acids, which are inserted in the outer membrane of the bacterium.
    - The nucleus a heteropolysaccharide derived from hexoses and heptoses.
    - Lipid A and the nucleus are bound by the sugar acid 2-keto-3-deoxyoctanate (KDO).

- The O chain is a repeating unit polymer of 1–8 glycosidic residues; this polymer is highly variable among bacterial species and genus.

In addition to the pyrogenicity of the endotoxin, an important role has been attributed to the adherence mechanism of the bacteria to the host cell; since in previous studies, it has been observed that when LPS is modified or not expressed, the adherence observed is modified or inhibited.

- Exotoxins: These are the macromolecules of protein origin, which are produced and later released to the medium by the microorganism. Depending on their mechanism of action, exotoxins are divided as follows:
  - Toxins Type I: These toxins modify the host's cells without internalizing in the cells; for example, the superantigens produced by Staphylococcus aureus and Streptococcus pyogenes.
  - Toxins Type II: Within this group there are hemolysins and phospholipases; this group of toxins is characterized by pore formation and/or destroying the membranes of the host cells. With this virulence factor, the pathogen can invade the host cell; for example, aerolysin and GCAT protein produced by Aeromonas spp.
  - Toxins Type III: These toxins are known as A/B due to their binary structure. Fraction B has the function of binding to the receptor of the cell and fraction A is the unit that possesses enzymatic activity, which, depending on the toxin and its mechanism of action, will be the damage to the cell; for example, the Shiga toxin produced by Escherichia coli O157:H7, the Cholera toxin (Ctx) produced by Vibrio cholerae, and the Anthrax toxin produced by Bacillus anthracis.

Exotoxins of Gram-negative enteropathogenic bacteria play an important role in the pathogenesis of diarrheal disease, causing hypersecretion of liquids without the destruction and death of intestinal mucosal cells. These toxins are generically referred to as enterotoxins that are different from cytotoxins.

There are also two other groups of toxins, those that alter the cytoskeleton and those with neurotoxic activity; however, some toxins may present activity corresponding to more than one of the groups described in table.

Toxin type	Definition
Enterotoxin	It produces a net secretion in ligated intestinal segments without histological evidence of intestinal lesion or damage to nonerythrocytic cells in in vitro tests. It stimulates the increase in the short circuit current (Isc) and the potential difference (PD) in the using chamber without evidence of intestinal damage; this result involves the secretion of (active) electrogenic anions. Additionally, a toxin can impair electrically neutral NaCl absorption, which also results in a net secretion of ions.

Cytoskeleton-altering toxin	It alters the cellular form and has been frequently shown to be caused by the F-actin rearrangement. The toxin can cause limited cell damage but is not lethal, and it may or may not be associated with the evidence of net secretion in in vivo or in vitro disease models in intestinal epithelial cells.
Cytotoxin	It causes cell or tissue damage, usually ending with cell death. The toxin may or may not be associated with net secretion in in vivo or in vitro disease models in intestinal epithelial cells.
Neurotoxins	It involves the release of one or more neurotransmitters from the enteric nervous system. It alters the activity of smooth muscle in the intestine.

Classification of Enteric Toxins

Toxins produced by pathogens involved in foodborne diseases are as follows:

Cholera toxin (Ctx) (Vibrio cholerae), Thermolabile toxin (LT) Thermostable toxin (ST) (Enterotoxigenic E. coli), Shiga Toxin (Shigella dysenteriae and E. coli O157:H7) Botulinum toxin (BTX) (Clostridium botulinum), CPE Enterotoxin (Clostridium perfringens), Alpha-Toxin, Beta-Toxin, Epsilon-Toxin and Iota-Toxin (C. perfringens), Toxin A/Toxin B (Clostridium difficile), Enterotoxins (A, B, C1, C2, D and E, G, H, I, J), Toxic Shock Syndrome Toxin (TSST-1), Cereulide, and hemolysin BL (HBL), nonhemolytic enterotoxin (NHE) (S. aureus), Citotoxin K or CytK (Bacillus cereus).

Risk Factors and Prevention Measures Associated with Food Poisoning

The main risk factor involved in bacterial food poisoning is food contamination by pathogenic bacteria that produce toxins; such contamination can occur at any time, that is, from the crop, in the case of vegetables or, just before eating them, due to the consumer's manipulation; in this way, all the people living on the earth are susceptible to food poisoning. Therefore, food poisoning is a worldwide public health problem, generally the most affected are children, the elderly, pregnant women, and immunocompromised people. As expected, individual factors such as age, gender, place of residence, socioeconomic factors, among others, are crucial in food poisoning acquisition and development.

Food contamination can occur from primary production to the final consumer, consequently, there are different contamination risks according to the practices carried out in the different stages such as agricultural, livestock, and fish production; industrialization (in the case of processed food); marketing (points of sale), and transportation to the final consumer (homes, community dining rooms, and restaurants).

During the primary production, producers should consider the particular characteristics of the environment where they grow or breed and reproduce livestock, by applying measures to prevent any pollution caused by the air, water, or natural fertilizers. In general, the main risk of contamination in primary production is the unsafe agricultural practices such as the use of manure as natural fertilizer and irrigation with sewage, which violates the fundamental principle of preventing, at all costs and contamination of raw materials from fecal matter.

Additionally, another important factor to ensure food safety and good quality is the adequate control of time and temperature when cooking, processing, cooling, and storing food. To achieve a good control of such parameters, it is necessary to consider the physical, chemical, and microbiological characteristics of each type of food, for example, water activity, pH and type, and the initial number of microorganisms presented there. Similarly, other aspects need to be taken into account such as shelf life and usage, that is, whether it is a raw, processed, packaged, or ready-to-eat food.

Microbiological contamination can occur through direct contact or through air, utensils, contact surfaces, or the handler's hands; therefore, ready-to-eat foods must be separated in space and time from raw or unprocessed foods. In addition, the latter must always be washed or disinfected. In all stages of the food chain, it is indispensable to use water; hence, this could be the main source of food contamination. It is then necessary to control and monitor the type and the source of the water used at each stage; however, when it is used for food handling, water has to be drinkable water that meets the physical, chemical, and microbiological criteria that its name requires.

In terms of facilities, it is important to establish and monitor systems that ensure their maintenance, cleaning, and sanitation. These systems also include an adequate waste management and an effective pest control. The latter constitute a potential risk of any type of contamination; that is why it is necessary to implement measures that prevent the entrance of any type of pests, as well as measures to avoid their nesting and proliferation. Finally, pest eradication must be carried out by any physical, chemical, or biological method that does not represent a threat to health and food safety.

Within the food chain, food transportation plays an important role in preventing contamination and proliferation of microorganisms in food; thus, it is necessary to consider measures to prevent any type of contamination and to provide an environment to control the proliferation of pathogenic microorganisms and the production of bacterial toxins. Some important factors to consider during food transportation are temperature, direct exposure to sunlight, humidity, and airflows. At this stage, the type of containers and the type of packaging also play an important role; the aforementioned and transport conditions should be chosen based on the characteristics of the food that is being transported.

Another important measure is the information that producers and suppliers offer to consumers regarding the characteristics and proper handling of prepackaged foods; this is why, generally, food must be packaged and labeled in such a way that the consumer has enough information to handle, store, and prepare the products appropriately without threatening his or her health. Labels should also include a batch number allowing rapid identification and market recalls of products potentially being dangerous for human consumption.

In general, microorganisms, more specifically bacteria, can proliferate under very different conditions; that is why they can be found in any type of environment. Even though bacteria are good at adapting to the environments they are in, there are certain conditions that promote bacterial growth more than others. These conditions include food, humidity, acidity, temperature, time, and oxygen; all of these are grouped in what is known as FATTOM (Food, Acidity, Time, Temperature, Oxygen, and Moisture). Knowing and avoiding these optimal conditions can help to prevent bacterial growth, bacterial infections, and food poisoning.

Most foods contain nutrients required for microbial growth, which makes them easy targets for the microorganisms to develop; therefore, perishable. To reduce the breakdown of food and to prevent foodborne diseases, the proliferation of microorganisms under certain conditions must be controlled, as well as the conditions that must be used to reduce food spoilage to lengthen the time during which physicochemical and organoleptic characteristics must be kept under minimum acceptance parameters. Factors affecting the proliferation rate of microorganisms can be considered as intrinsic and extrinsic.

Intrinsic Parameters

Intrinsic factors affecting the proliferation rate are more related to the internal characteristics of food products, and the way in which these characteristics maintain or affect the growth of microorganisms; these factors include water activity, pH, oxidation-reduction potential, content and type of nutrients, inhibiting substances, and biological structures.

Water Activity

It is defined as the amount of water available for the growth of microorganisms; microbial proliferation decreases when water availability also decreases. The water available for metabolic activity determines the degree of microbial growth instead of the total moisture content. The unit of measurement for the water that microorganisms require is usually expressed as water activity (A_w), which is defined as the water vapor pressure of food substrate, divided by the water vapor pressure of pure water, at the same temperature. This concept is related to relative humidity (RH), thus: $RH = 100 \times A_w$. The approximate optimal A_w for the growth of most microorganisms is 0.99; most bacteria require an A_w greater than 0.91 to grow. Gram-negative bacteria require higher values than Gram-positive bacteria. Most of the natural food products have an A_w of 0.99 or more. Generally, bacteria have the highest requirements of water activity, fungi have the lowest, and yeasts have intermediate requirements. Most bacteria that decompose food do not grow with an A_w less than 0.91, but fungi and yeasts can grow with values of 0.80 or less, including surfaces partially dehydrated. The lowest value reported for bacteria in food is 0.75 for halophytes, while xerophilic fungi and osmophilic yeasts have shown growth at A_w values of 0.65 and 0.61, respectively.

pH

The pH is defined as the negative logarithm of hydronium ions concentration; it is considered as a unit of measure to establish acidity or alkalinity levels of a substance, in this case food, and it is determined by the number of free hydrogen ions (H^+). The effects of adverse pH affect at least two aspects of the microbial cell-functioning of its enzymes and nutrients transportation to the cell.

The cytoplasmic membrane of microorganisms is relatively impermeable to H^+ and OH^- ions; its concentration in the cytoplasm remains reasonably constant, despite the wide variations that may occur in the pH of the surrounding medium. When microorganisms are in an environment below or above the neutral level, their ability to proliferate depends on their ability to change the environmental pH to a more appropriate range, since key components like DNA or ATP require a neutral medium.

The pH for the optimal growth of most microorganisms is close to neutrality (pH = 6.6–7.5). Yeasts can grow in an acid environment and thrive in an intermediate range (4.0–4.5), although they survive in values between 1.5 and 8.5. Fungi tolerate a wide range (0.5–11.0), but their growth is generally higher in an acid pH (too acid for bacteria and yeast). Bacterial growth is usually favored by pH values closer to the neutral level. Nevertheless, acidophilic bacteria grow on substrates with a pH of up to 5.2 and below that point the growth reduces dramatically.

In general, fruits, vinegars, and wines have pH values lower than those required for bacterial growth, so they can usually be decomposed by fungi and yeasts. Most vegetables have pH values lower than those from fruits, and consequently, vegetables are more exposed to bacterial or fungi decomposition. In contrast, most meats and sea products have pH values equal or greater than 5.6, making them susceptible to decomposition by bacteria, fungi, and yeasts.

Oxidation-reduction Potential

The oxidation-reduction potential (O/R) is an indicator of the oxidizing and reducing power of a substrate; that is, the O/R potential of a substrate can be generally defined as the ease with which a substrate loses or gains electrons (when a food product loses electrons, it oxidizes, whereas, when it gains electrons it is reduced; thus, a food product that easily gives electrons is a good reducing agent and the one that receives electrons is a good oxidizing agent). To achieve optimum growth, some microorganisms require reducing conditions and others require oxidizing conditions. The O/R potential of a system is expressed with the Eh symbol (when electrons are transferred from one compound to another, a potential difference is created between the two compounds; this difference can be measured and expressed as millivolts [mV]). The more oxidized a substance is, the more positive the electrical potential will be; and the more reduced a substance is, the more negative the electrical potential will be. When the concentration of oxidant and reducer is equal, there is an electrical potential of zero.

Saprophytes that are capable of transferring hydrogen as H^+ and e^- (electrons) to molecular oxygen arc acrobic; that is, aerobic microorganisms require positive Eh values (oxidized) for their growth, whereas anaerobic microorganisms require negative values of Eh (reduced). Facultative microorganisms can grow under any of the conditions. It has to be considered that maximum and minimum Eh values (in mV) necessary for aerobic and anaerobic growth could be lethal to the other group. Among food substances that help to maintain reducing conditions are the –SH groups in meats and the ascorbic acid, as well as, reducing sugars in fruits and vegetables. Some aerobic bacteria grow better under slightly reducing conditions being known as microaerophiles such as Lactobacillus and Campylobacter. Most of fungi and yeasts found in food are aerobic, although a few tend to be facultative anaerobes. Regarding the Eh value of food, vegetables, especially juices, tend to have Eh values of +300 to +400 mV; so, it is not surprising to find that aerobic bacteria and fungi are the common cause of decomposition in this type of products. Meats have Eh values around −200 mV; in ground meats, Eh is usually around +200 mV. Various types of cheese show Eh values between −20 and −200 mV.

Content of Nutrients

Microorganisms have nutritional requirements, most of them need external sources of nitrogen, energy, minerals, as well as vitamins, and related growth factors; these requirements are found in our food, so if they have the right conditions to develop, they will. In general, fungi have the lowest nutrient requirement, followed by Gram-negative bacteria, then yeasts and finally, Gram-positive bacteria, which have the highest requirements.

The primary sources of nitrogen used by heterotrophic microorganisms are amino acids. A great number of other nitrogen compounds may serve for this function for several types of organisms. For example, some of them can use free nucleotides and amino acids, while others can be capable of using peptides and proteins. In general, simple compounds like amino acids will be used by almost all of the organisms before attacking more complex compounds such as high molecular weight proteins. The same applies to polysaccharides and lipids.

Microorganisms in food tend to use as energy sources, sugars, alcohols, and amino acids. Fungi are the most efficient in the use of proteins, complex carbohydrates, and lipids because they contain enzymes capable of hydrolyzing these molecules into simpler components; many bacteria have a similar capacity, but most yeasts require simpler molecules. All microorganisms need minerals, although vitamin requirements vary. Fungi and some bacteria can synthesize enough B vitamins to meet their needs, while others need to have a source of vitamins, food products being an excellent source of them.

Gram-positive bacteria are the ones that have lower synthesized capacity, so they need one or more of these components to grow. In contrast, Gram-negative bacteria and fungi are capable of synthesizing the most, if not all, of their requirements and consequently, these two groups of organisms can grow in food products with low content of B vitamins.

Extrinsic Parameters

Food factors are very important for the development of microorganisms; there are external or extrinsic factors. This term refers to environmental factors that affect the growth rate of microorganisms; these factors include temperature, oxygen availability, and relative humidity, as well as, the presence and activities of other microorganisms.

Storage Temperature

Microorganisms have an optimal range, as well as a minimum and maximum temperature to grow. Therefore, ambient temperature determines not only the proliferation rate, but also the genera of microorganisms that are going to be developed, along with the microbial activity degree that is registered. The change in only a few degrees in temperature will favor the growth of completely different organisms, and it will result in a different type of food decomposition and foodborne disease. Due to these characteristics, thermal treatment is employed as a method to control microbial activity.

The optimal temperature for the proliferation of most microorganisms ranges from 14 to 40 °C, although some genera develop below 0 °C, and other genera grow at temperatures above 100 °C. Nevertheless, food quality must be taken into account when selecting storage temperature. Although it can be desirable to storage all food products at temperatures equal or less to those of refrigeration, this is not the best thing to do to maintain a desirable quality in some food products such as banana, whose quality is best maintained in storage at 13–17 °C than at 5–7 °C. Similarly, many vegetables are favored at temperatures near 10 °C such as potatoes, celery, cabbage, and many others. In each case, the success of storage temperature depends, to a large extent, on the relative humidity and the presence or absence of gases such as carbon dioxide and ozone.

Oxygen Availability and Presence of other Gases in the Environment

Like temperature, the oxygen availability determines the microorganisms that will be active. Some have an absolute requirement for oxygen, while others grow in total absence of it, and others may grow with or without oxygen. Microorganisms that require free oxygen are called aerobic microorganisms, while those that thrive in the absence of oxygen are called anaerobic; and those that grow both in presence or absence of free oxygen are known as facultative microorganisms.

Carbon dioxide is the most important atmospheric gas that is used to control food microorganisms. Along with oxygen, it is used in packaged food with modified atmosphere. Ozone is another atmospheric gas with antimicrobial properties, and for decades, it has been used as an agent to lengthen shelf life of certain types of food. Although being effective against a variety of microorganisms, it is a highly oxidizing agent;thus, it cannot be used in food products with high lipid content, as it could accelerate rancidity. Normally, ozone levels of 0.15–5.00 ppm in the air inhibit the growth of some bacteria that decompose food as well as yeast growth.

Relative Humidity in the Environment

Relative humidity (RH) of the environment is important from the point of view of water activity within food and the growth of microorganisms on surfaces. This extrinsic factor affects microbial growth and can be influenced by temperature. All microorganisms have a high-water requirement, this being needed for their growth and activity.

When the A_w of a food product is set at 0.60, it is important that this food is stored under RH conditions that do not allow food to draw humidity from the air and, therefore, it increases its own A_w from the surface and subsurface to an extent where microbial growth can occur. A high relative humidity can cause humidity condensation in food, equipment, walls, and ceilings. Condensation causes wet surfaces, which lead to microbial growth and decomposition. Microbial growth is inhibited by a low relative humidity. When food products with low A_w values are placed in high RH environments, food takes in moisture until they reach balance. Similarly, food products with high A_w lose moisture when placed in an environment with low RH. There is a relationship between RH and temperature that must be taken into account when selecting the appropriate storage environments for food products. Overall, the higher the temperature, the less the RH, and vice versa.

Bacteria require higher humidity than yeasts and fungi. The optimal relative humidity for bacteria is 92% or higher, while yeasts prefer 90% or higher, and fungi thrive if the relative humidity is between 85 and 90%. Food products suffering superficial decomposition by fungi, yeasts, and specific bacteria, should be stored under low RH conditions. Poorly packed meats such as whole chickens and beef cuts, tend to suffer a lot of superficial decomposition inside the refrigerator before internal decomposition occurs, usually, due to high RH in refrigerators, and to the fact that the biota decomposing meat is essentially aerobic in nature.

Although it is possible to decrease the possibility of superficial decomposition in certain food products by storing them in low RH conditions, it should be remembered that the food itself will lose moisture into the atmosphere under such conditions, and thus, it will become undesirable. When selecting appropriate RH conditions, there should be taken into account both the possibility of superficial microbial growth and the quality that the food product needs to have. By altering the gas atmosphere, it is possible to delay superficial decomposition without lowering the relative humidity.

Presence and Activities of other Microorganisms

Some food origin organisms produce substances that can inhibit or be lethal for other organisms; these include antibiotics, bacteriocins, hydrogen peroxide, and organic acids. Bacteriocins produced by lactic acid-producing bacteria originated in various food products such as meat, are of high interest. Bacteriocins produced by Gram-positive bacteria are biologically active proteins with bactericidal action. Some bacteriocins produced by these bacteria inhibit a variety of food pathogens including, B. cereus, C. perfringens, Listeria spp., A. hydrophila, and S. aureus, among others.

Normally food products can reach the final consumer at home, in community dining rooms, or restaurants. Measures to prevent food poisoning should be implemented at these locations, particularly in areas where large volumes of food are distributed such as cold chain, frozen chain, hot chain, and vacuum cooking. Likewise, in the frozen chain, food temperature is gradually lowered to −18 °C and defrosted at temperatures higher than 65 °C at the time it will be served to the costumer (not before); while in the hot chain, for example, in a buffet, food is kept at temperatures higher than 65 °C and it should be consumed within 12 h maximum.

Other important measures are the use of food preservation methods, which can be physical or chemical. Within the physical methods, there are the traditional or industrial pasteurization, dehydration, preservation in modified atmosphere, and irradiation. In order to maintain an adequate quality control and to minimize the risk of food poisoning, microbial markers can be used; these markers do not represent a potential health risk, however, a large number of them indicate deficiencies in hygiene and sanitary quality of food products; it also leads to a decrease in the shelf-life and could be related to the presence of pathogenic microorganisms. The main microbial markers are aerobic mesophilic, total coliforms, fecal coliforms, Enterococci, E. coli, S. aureus, and lactic acid bacteria.

Once the risk factors are identified, it is necessary to establish a system that allows to prevent and decrease all of them; to do this, a method with scientific basis and systematic profile has been established, this is known as Hazard Analysis and Critical Control Point (HACCP). A microbiological approach should consider the type of microorganism or metabolite (toxins) that threatens human health; the analytical methods for its detection and quantification; the number of samples to be taken and the size of the analytical unit; and the microbiological limits considered to be adequate at specific points in the food chain.

Foodborne Diseases

In food products, we can find different types of toxins such as, bacterial, fungal (mycotoxins), algae or plant toxins, as well as metals, toxic chemicals (zinc, copper, and pesticides), and physical contaminants that can cause diseases in people who eat them; all of these can cause the well-known "foodborne diseases".

Foodborne diseases can be classified into two groups: poisoning and infection.

- Poisoning is caused by the intake of chemical or biological toxins; or toxins produced by pathogens, the latter can be found in food, even if the bacterium is not there.
- Infection is caused by the intake of food containing viable pathogens. Furthermore, a toxic infection (toxicoinfection), formerly known as a toxin-mediated infection, is caused by eating food with bacteria that grow and produce a toxin inside the body.

To meet the ideal conditions, microorganisms in food grow and produce toxins. By ingesting contaminated food, toxins are absorbed through the intestinal epithelial lining, and it causes local tissue damage. In some cases, toxins can reach organs such as the kidney or the liver, the central nervous system or the peripheral nervous system, where they can cause some damage.

The most common clinical symptoms of foodborne diseases are diarrhea, vomit, abdominal cramps, headaches, nausea, pain, fever, vomit, diarrhea with mucus and blood (dysentery), and rectal tenesmus. Some of the microorganisms causing foodborne diseases, either from poisoning, intoxication or toxicoinfection are described in Tables. These diseases are generally diagnosed based on the patient's clinical record or their symptoms.

Pathogens that cause Infection

Bacteria	Disease/medical complications	Food products involved
Salmonella enterica serovar Typhi and Salmonella enterica serovar Paratyphi	Typhoid and paratyphoid fever.	Undercooked pork, beef and poultry, contaminated eggs, and milk.
Salmonella spp.	Salmonellosis (Salmonella Typhimurium, Salmonella Enteritidis).	Undercooked poultry, cauliflowers, and tomatoes.
Vibrio vulnificus	Septicemia in people with underlying diseases or people who are taking immunosuppressive drugs or steroids.	Seafood, usually oysters.
Mycobacterium bovis	Cervical lymphadenopathy, intestinal lesions, chronic cutaneous tuberculosis.	Contaminated milk.
Mycobacterium avium, subspecies paratuberculosis	Crohn's disease.	Pasteurized milk.
Listeria monocytogenes	Meningitis, encephalitis, sepsis in pregnant women, intrauterine or cervical infection that can lead to miscarriage or birth of a dead child.	Raw beef, pork, poultry, vegetables and milk, cheese, ice cream, smoked fish, and raw fish.

Pathogens that cause Intoxication

Bacteria	Disease/medical complications	Food products involved
Clostridium botulinum	Paralysis of arms, legs, trunk, and respiratory muscles.	Mixture of oil and nonacid garlic, potatoes cooked at high temperatures, and stews.
Bacillus cereus	Fried rice syndrome.	Rice cooked at high temperatures, sauces, soups, and puddings.
Staphylococcus aureus	Toxic shock syndrome.	Meat and meat products cooked at high temperatures, poultry, and salads with mayonnaise.

Pathogens that cause Toxico-infection

Bacteria	Disease/medical complications	Food products involved
Escherichia coli O157:H7	Hemorrhagic colitis, Hemolytic uremic syndrome in children.	Hamburgers, nonpasteurized milk, contaminated water, spinach, and lettuce.
Shigella spp.	Hemolytic Uremic Syndrome.	Salads, lettuce, raw vegetables, and milk.
Aeromonas spp.	Meningitis, peritonitis, myocarditis, hemolytic uremic syndrome, necrotizing fasciitis in wounds.	Meat (beef, sheep, pork and chicken), vegetables, eggs, fish, seafood, and prepared food.
Cronobacter sakazakii	Permanent neurological or developmental deficits; death.	Powdered infant formula.
Vibrio parahaemolyticus	Gastroenteritis, septicemia and wound infection. Severe infections in immunocompromised people.	Raw or undercooked seafood, usually oysters.
Clostridium perfringens	Clostridial necrotizing enteritis.	Meat juice, stews, cooked beans, meat cooked at high temperatures.
Campylobacter spp.	Campylobacteriosis, arthritis, meningitis, Campylobacter jejuni can cause Guillain-Barré Syndrome.	Cheese made with raw milk and chicken meat.
Yersinia enterocolitica	Yersiniosis, enterocolitis, pseudoappendicitis, mesenteric lymphadenitis, infections in wounds, joints and the urinary tract, and Reiter's syndrome.	Nonpasteurized milk, tofu, nonchlorinated water, undercooked meat, oysters and fish.
Vibrio cholerae serogroup O1 or serogroup O139	Cholera.	Contaminated water and raw seafood.
Vibrio cholerae serogroup no-O1	Less severe than Cholera; gastrointestinal infections, sepsis.	Raw, semicooked or recontaminated fish and shellfish after cooking.

Foodborne Diseases caused by Bacterial Toxins

This topice will be addressed to some diseases caused by consuming food contaminated with bacterial toxins or microorganisms that produce them. Among some of the most important diseases are the ones transmitted by V. cholerae, S. aureus, B. cereus C. perfringens, C. botulinum and Listeria monocytogenes.

Vibrio Cholerae

V. cholerae has a free life cycle, it is ubiquitous in aquatic environments; it is able to remain virulent without multiplying in fresh water and sea water for a long time. They are more frequent in temperate waters and can be isolated in seafood and fish. The most notable species are V. cholerae O1 and O139, causative serogroups of Cholera. Non-O1 strains and the rest of the species cause cholera-like diarrheal syndromes, but they are not as severe, although they frequently produce extraintestinal infections.

The CTX toxin (Cholera toxin) is the main virulence factor of V. cholerae O1 (Ogawa, Inaba, and Hikojima serotypes, Classical and El Tor biotypes) and O139; it contributes to cause profuse diarrhea, after an incubation period from 2 h to 5 days; stools have the appearance of rice water, there is dehydration and electrolyte imbalance, which can lead to death. Approximately 75% of the infected people are asymptomatic, that is, they do not develop the symptoms aforementioned; however, the pathogen is shed in their feces for 7–14 days, which is a very serious source of contamination since it is possible to infect others. The most vulnerable groups are children, adults, and people infected with the HIV virus.

This toxin can be identified by the presence of the ctxAB gene. V. cholerae no-O1 has the ctx gene but it is rarely expressed; nevertheless, a faster test is not yet available, although the WHO is currently in the process of validating new rapid diagnoses. The bacteria can be isolated and identified from stool samples by using laboratory procedures.

Efficient treatment resides in prompt rehydration through oral solutions or intravenous fluids. The use of antibiotics is suggested only when there is severe dehydration. The supply of safe drinking water, the adequate sanitation, and food security are essential to prevent the emergence of Cholera. Moreover, vaccines administration has emerged because control measures to prevent contamination are insufficient; this is the reason why oral vaccines have been developed as tools to prevent outbreaks. These vaccines are given to more vulnerable populations in areas where the disease is endemic. Experience in different mass vaccination campaigns in countries such as Mozambique, Indonesia, Sudan, and Zanzibar clearly indicates that vaccination requires careful and early planning and preparation, and therefore, it cannot be improvised at the last minute.

The lack of toxicity combined with stability and the relative ease to express the Cholera Toxin Subunit B (CTB) has contributed to be an easily manageable adjuvant. The ability to express protein in a wide variety of organisms broadens even further its application potential. CTB is currently being used in vaccines such as Dukoral, a vaccine against V. cholerae that consists of dead bacteria and recombinant CTB. It has been approved as adjuvant for vaccines in Europe and in Canada; and given the excellent adjuvant effect, this protein is likely to play an important role in vaccine formulation in the future.

Staphylococcus Aureus

Staphylococcal foodborne illness is one of the most common diseases acquired by

S. aureus. It is one of the most concerned diseases by public health programs in the world; it is due to the production of one or more toxins by the bacteria during their growth at permissive temperatures; however, the incubation period of the disease depends on the amount of ingested toxin. Small doses of enterotoxins can cause the disease; for example, a concentration of 0.5 ng/mL in contaminated chocolate milk has been reported to cause large outbreaks.

S. aureus produces various toxins. Staphylococcal enterotoxins are a family of nine thermostable enterotoxin serotypes belonging to a large family of pyrogenic toxins (superantigens). Pyrogenic toxins can cause immunosuppression and nonspecific T cell proliferation. Enterotoxins are highly stable and they resist high temperatures (which makes them suitable for industrial use) and environmental conditions of drying and freezing. They are also resistant to proteolytic enzymes (pepsin and trypsin) at low pH, enabling them to be fully functional in the digestive tract after infection.

The mechanism by which poisoning is caused is not entirely clear yet. However, enterotoxins have been observed to directly affect the intestinal epithelium and the vagus nerve causing stimulation of the emetic center. It is estimated that 0.1 μg of enterotoxin can cause staphylococcal poisoning in humans. Apart from causing poisoning, S. aureus can also cause toxic shock syndrome due to the production of the Toxic Shock Syndrome Toxin 1 (TSST-1) and Enterotoxin Type B.

Symptoms include nausea, vomit, abdominal cramps, salivation, diarrhea could be present or absent. The first three symptoms are the most common ones. Usually, it is a self-limiting disease and can be cured in 24–48 h, but it can become severe, especially in children, the elderly, and immunocompromised people. Toxic shock syndrome is characterized by high fever, hypotension, erythematous rash (similar to scarlet fever, peeling of the skin during recovery, flu-like symptoms, vomiting, and diarrhea).

The diagnosis of the disease is carried out by detecting the staphylococcal enterotoxin in the food or by recovering at least 10^5 S. aureus/g from food leftovers. The enterotoxin can be detected by several methods: bioassays, molecular biology, and immunological techniques. The isolated strains can be genetically characterized by multilocus sequences from the spa or SCCmec gene, and pulsed-field electrophoresis.

The mainly involved food products in outbreaks and where S. aureus can grow optimally, since they are stored at room temperature, are meat and its derived products, poultry and eggs, milk and its derived products, salads, and bakery products (cream-filled cakes and stuffed sandwiches).

Other factors that must be taken into account are the emergence of methicillin resistant strains, which may be found in food (mainly in meat and milk). It is important to note that many of the isolates obtained from outbreaks are not tested for antimicrobial susceptibility; due to the various problems that these strains can create, the antimicrobial susceptibility test should be performed. They have been reported to be causative agents of outbreaks in blood infections and wounds in immunocompromised patients in hospitals.

Foodborne illness due to S. aureus may be preventable. It is known that the permissible temperature for the growth and production of the enzyme is between 6 and 46 °C; thus, food products could be cooked above 60 °C and refrigerated below 5 °C. Therefore, maintaining the cold chain of food can prevent the growth of the microorganism. By using good manufacturing practices and good hygiene practices, the contamination by S. aureus can be prevented.

Bacillus Cereus

B. cereus is a ubiquitous microorganism in the environment, and it can easily contaminate any food production and processing system, due to the formation of endospores. The bacterium can survive pasteurization and cooking processes.

It has been demonstrated that this microorganism produces, cereulide or emetic toxin; three enterotoxins, hemolysin BL (HBL), nonhemolytic (NHE), cytotoxin K (CytK), which are responsible for the emetic syndrome and diarrhea; and three phospholipases, phosphatidylinositol hydrolase, phosphatidylcholine hydrolase, and hemolytic sphingomyelinase. Cereulide is a thermostable cyclic peptide that causes emesis by stimulating the afferent vagal pathway through its bond to the serotonin receptor. The toxin is produced during the stationary phase of growth of the microorganism and it accumulates in food over time. The structure of the toxin explains its resistance to food processing methods. In contrast, inside the small intestine of the host, the thermolabile enterotoxins, HBL and NHE, produced during the exponential phase of the vegetative growth of the bacterium are the cause of diarrheal syndrome; the proteins that form enterotoxins (binding and lithic factors) are unable to traverse intact the gastric barrier; that is why it is considered that preformed or extracellular enterotoxins in food are not involved in the pathogenesis of the bacterium. It is believed that the spore germination that reaches the small intestine, the growth, and the simultaneous production of the enterotoxin are the ones that cause diarrhea. HBL is a hemolysin formed by three components, two protein subunits (L2 and L1), and one B protein; it has hemolytic, cytotoxic, and dermonecrotic effect, and it induces vascular permeability. NHE also consists of three components: NheA, NheB, and NheC. It has been demonstrated that strains producing emetic toxin do not produce enterotoxin. The cytotoxin K is similar to the Alpha-toxin of S. aureus and the Beta-toxin of C. perfringens.

Furthermore, the enterotoxin FM (EntFM) has been described; it is a 45 kDa polypeptide encoded by the entFM gene, located in the bacterial chromosome. It has not been directly involved in food poisoning; however, the presence of the gene in strains that cause diarrheal outbreaks has been detected; in experiments with mice and rabbits, it causes vascular permeability.

The emetic syndrome is characterized by nausea and vomit similar to those produced by S. aureus poisoning. Symptoms appear soon after consuming food contaminated with the preformed toxin. Generally, poisoning develops with mild symptoms, usually lasting no more than 1 day, but severe cases require hospitalization. The diarrhea that is caused belongs to the secretory type, similar to the one produced by V. cholerae. Colic pain occurs similar to that of C. perfringens poisoning. Both syndromes are self-limiting.

Enterotoxins can be detected by immunoassays or molecular biology (conventional PCR and multiple PCR) by looking for the ces gene (nonribosomal production of cereulide); by detecting the hblD, hblC, and hblA genes encoding the L1, L2, and B protein components of the HBL toxin, respectively; or the nheA, nheB, and nheC genes of the NHE toxin components. The 16S ribosomal gene can be looked for by real-time PCR.

Apart from causing food poisoning, B. cereus can also cause local and systemic infections in immunocompromised patients, neonates, people taking drugs, and patients with surgical or traumatic wounds, or catheters.

The most susceptible food products to be contaminated include flours, meats, milk, cheese, vegetables, fish, rice and its derived products; generally, in food with high content of starch. The strains produced by the emetic toxin grow well in rice dishes (fried and cooked) and other starchy products; although, there have been studies where it has been demonstrated that the toxin can be in different types of food products; while strains producing diarrheagenic toxins grow in a wide variety of food products, from vegetables to sauces and stews.

Strains isolated from infections have been shown to be sensitive to chloramphenicol, clindamycin, vancomycin, gentamicin, streptomycin, and erythromycin; they are resistant to β-lactam antibiotics, including third-generation cephalosporins.

Inadequate cooking temperatures, contaminated equipment, and poor hygiene conditions at the food processing and preparation sites are the major factors that contribute to food poisoning by B. cereus and its toxins; that is why, it is suggested to store food at temperatures lower than 4°C or to cook them at temperatures higher than 100°C, and to reheat or cool food rapidly, to avoid prolonged exposure to temperatures that allow spore germination and to diminish the risks of a possible poisoning.

Clostridium Perfringens

C. perfringens is an anaerobic bacterium that creates spores that survive in soil, sediments, and areas subject to both human and animal fecal contamination. It is widely distributed in the environment and is frequently found in the human intestine and in several domestic and wild animals' intestines.

C. perfringens is classified into five groups (A, B, C, D, and E), due to the different toxins it produces (alpha, beta, epsilon, and iota). The Alpha-toxin is produced by all the five groups. The Beta-toxin forms selective pores for monovalent ions in the lipid bilayers, functioning as a neurotoxin capable of producing arterial constriction. The Epsilon-toxin is the most potent clostridial toxin after tetanus and botulinum neurotoxins (BoNTs). It is produced and secreted by a prototoxin that acquires its maximum biological activity by undergoing a specific proteolytic cleavage; its activation can be catalyzed by trypsin, chymotrypsin, and a zinc metalloprotease.

The toxin receptor is unknown, but it is known to be a surface protein anchored by glycosylphosphatidylinositol. Its main biological activity is the edema generation; it is lethal but not hemolytic. The Iota-toxin is a member of the binary toxin family, since it is formed by a binding peptide (Ib) necessary for the internalization of the enzymatic peptide (Ia; ADP-ribosyltransferase). Proteolytic removal of a propeptide fragment is required to allow Ib to be inserted into the membrane and to interact with Ia. Ib, when inserted into the membrane, forms a heptameric pore that allows the exit of K^+ and Na^+ ions, and the entry of Ia, which once inside the cell, is ribosylated by the G-actin; it depolymerizes the filaments of Actin by destroying the cellular cytoskeleton. The Iota-toxin is dermonecrotic, cytotoxic, enterotoxic, and induces intestinal histopathological damage.

However, the virulence of this bacterium is not only due to the presence of these 4 toxins; there have also been described 15 toxins within which the CPE enterotoxin is responsible for causing diarrhea in humans and animals, and it is produced by Type A strains. This toxin is associated with 5 or 15% of gastrointestinal diseases in humans different from food poisoning such as diarrhea produced by antibiotics; the NetB toxin is frequently related to necrotic enteritis in birds and the Beta2-toxin is apparently associated with enteritis. The production of toxins in the digestive tract is associated with sporulation. The disease is foodborne; and only one case has implied the possibility of poisoning caused by the preformed toxin.

C. perfringens causes food poisoning characterized by severe abdominal cramps and diarrhea beginning after 8–22 h of food intake, the disease ends 24 h after the intake; although, in some cases the disease may persist for 1–2 weeks. Additionally, there is a more severe but less frequent disease caused by eating a food product contaminated with type C strains; this disease is known as necrotic enteritis or pig-bel disease, and it is often fatal. Deaths caused by necrotic enteritis are due to intestinal infection and necrosis, as well as by septicemia, the elderly people being the most affected population.

The disease diagnosis is confirmed by the presence of the toxin in the stools of patients; either by traditional methods (culture from the stools or the food involved) or by molecular methods by looking for the following genes: cpe (CPE toxin), plc (Alpha-toxin), and etx (Epsilon-toxin).

Among the main food products involved are meat and its derived products. The disease can be prevented if the food has been properly cooked; although, there may be a risk of cross-contamination if the cooked food comes in contact with raw and contaminated ingredients, as well as contaminated surfaces.

There is no specific treatment or established cure for the infections caused by the toxins of the bacteria. Supportive care includes administration of intravenous fluids, oral rehydration salts solutions, and medication for fever and pain control. The treatment of gas gangrene is based on surgical measures with debridement and removal of the affected tissue and administration of high doses of antibiotics. Necrotizing enterocolitis is treated systemically with penicillin G, metronidazole or chloramphenicol; 50% of the cases require surgical treatment in which a segmental jejunum resection is performed. The antibiotics active against anaerobic bacteria are effective; however, there are strains resistant to penicillin and clindamycin, therefore, it is suggested to perform antimicrobial susceptibility tests, especially in patients with severe disease and those requiring long-term treatments.

Clostridium Botulinum

C. botulinum is a spore-forming microorganism; these spores can remain viable for long periods of time when the environmental conditions are absolutely unfavorable for the development of the microorganism.

Four groups are recognized in C. botulinum, as well as seven antigenic variants of botulinum neurotoxins (A–G). Groups I and II are primarily responsible for botulism in humans; Group III is responsible for causing botulism in several animal species, and Group IV appears not to be associated with the disease in either humans or animals. Group I is also known as C. botulinum-proteolytic (mesophilic microorganisms), while group II is known as C. botulinum-non-proteolytic (psychrophilic microorganisms). Group I forms spores that are highly resistant to heat, the "Botulinum cook" (121 °C/3 min) given to canned foods with a low content of acid is designed to inactivate them; neurotoxins formed in this group are A, B, F, and H. Group II forms moderately heat-resistant spores, and the neurotoxins formed are B, E, and F. Botulism types A, B, E, and F rarely cause the disease in humans, whereas in animals it is caused by types C and D. Toxins are resistant to proteolytic reactions and to denaturation into the gastric apparatus. Botulinum toxins are metalloproteins with endopeptidase activity that require zinc; the general structure shows two chains with a molecular weight of 150 kDa, the double chain is subdivided into a heavy (H) structure constituted by a nitrogen terminal domain (HN), and a carboxyl-terminal (HC), and a lighter structure (L) that performs the catalytic function of the toxin. HC is responsible for binding to presynaptic receptors for internalization, and HN is called translocation domain.

C. botulinum, is a bacterial species known simply for producing the botulinum toxin. The number of genes in Group II strains coding for the neurotoxin is variable; there may be one to three genes that encode one to three different neurotoxins; if there are two genes, there can be one active toxin and an inactive toxin, or both toxins can be active. In Group II, the presence of only one gene has been described, that is why there is only one neurotoxin; however, in other studies it has been demonstrated that in Type F strains the toxin has part of Type B and Type E neurotoxins. Botulinum neurotoxins form complexes with accessory proteins (hemagglutinin and non-hemagglutinin), which protect the neurotoxin and facilitate their adsorption into the host. The hemagglutinin complex of the neurotoxin type A specifically binds the cell adhesion protein, E-cadherin, by binding the epithelial cell and facilitating the adsorption of the neurotoxin complex from the intestinal lumen. Dual toxin-producing strains have been isolated from botulism in humans, the environment, and food; recently there have been found strains that produce three botulinum toxins called F4, F5, and A2. The significance of producing two or more toxins on virulence, as well as the evolutionary consequences are not yet clear. Phylogenetic studies show evidence of horizontal gene transfer; the production of the dual toxin in Group I and the production of a single toxin in Group II is still not clear. Therefore, studies with toxins isolated and purified from the different groups of C. botulinum are still being carried out.

Botulism is a severe disease with a high fatality rate. The typical symptoms are flaccid muscle paralysis, sometimes it starts with blurred vision followed by an acute symmetrical decrease of bilateral paralysis that, if untreated, can lead to paralysis of the respiratory and cardiac muscles. If severe cases are not fatal, the patient may improve his/her condition after months or even years. There are three types of botulism: infant/adult intestinal botulism, wound botulism, and foodborne botulism. The first type (infant/adult intestinal botulism) is an infection associated with the multiplication of the microorganism and neurotoxin formation in the intestine; the second type (wound botulism) is an infection associated with cell multiplication and toxin formation in the wound, often acquired after drug abuse; and the third type (foodborne botulism) is a poisoning caused by the consumption of neurotoxin preformed in food. An amount of 30 ng of toxin is enough to cause the disease and sometimes death. Symptoms appear between 2 h and 8 days after the intake of contaminated food, although they may occasionally appear between 12 and 72 h.

Botulism can be diagnosed only by clinical symptoms, but its differentiation from other diseases can be difficult. The most effective and direct way of confirming the disease in the laboratory is by demonstrating the presence of the toxin in the serum, in stools of patients, or in food products consumed by them. One of the most sensitive and widely used methods to detect the toxin is through neutralization in a rodent. This test takes 48 h, and culture of specimens takes from 5 to 7 days. Infant botulism is diagnosed by detecting botulinum toxins and the microorganism in the stools of children.

Approximately 90% of the reported cases are related to the consumption of home-made preserved food, especially vegetables; the industrial preparation of meat and fish is rarely associated with botulism. Food products where spores of the bacteria or the botulinum toxin can be found are canned corn, pepper, soups, beets, asparagus, ripe olives, spinach, tuna chicken, chicken liver, ham, sausages, stuffed eggplants, lobster, and honey, just to name a few.

To prevent the chances of getting botulism through food, it is necessary to carry out appropriate control measures in food processing and handling, especially when new technologies are introduced or modified. Applying the "Botulinum cook" in the modern industry allows to secure canned foods. The use of chlorine and chlorinated compounds can help sanitize places that handle food industrially. Spores can also be inactivated with ozone and ethylene oxide.

Listeria Monocytogenes

L. monocytogenes is a facultative intracellular microorganism widely distributed in nature, capable of surviving both in the soil and the cytosol of a eukaryotic cell. Considering somatic (O) and flagellar (H) antigens, this bacterium can be classified into 13 serotypes (1/2a, 1/2b, 1/2c, 3a, 3b, 3c, 4a, 4b, 4b, 4c, 4d, 4e, 7), but only the serotypes 1/2a, 1/2b, and 4b are responsible for more than 98% of the cases of human listeriosis. Furthermore, it has also been grouped into four lineages (I, II, III, and IV), where lineage I (serotypes: 1/2b, 3b, and 4b) and lineage II (serotypes: 1/2a, 1/2c, 3a, and 3c) include most strains isolated from clinical cases; lineage I strains have a greater pathogenic potential. Lineages III and IV include strains of serotypes 4a, 4c, and an atypical 4b.

L. monocytogenes expresses multiple virulence factors, which allow to enter and survive in several nonphagocytic cells. After cellular internalization, listeriolysin O (LLO) and two phospholipases mediate the escape of the bacterium from the endocytic vesicle into the cytoplasm, where the microorganism divides and submits the F-actin based on mobility to spread from cell to cell. The LLO (coded by the gene hly) is a cholesterol-dependent toxin; it is able to form pores in the membrane of phagosomes, allowing L. monocytogenes to escape from primary and secondary vacuoles. The cytolytic activity of LLO increases with the action of a phosphatidylinositol phospholipase C (PI-PLC), the substrate of which is phosphatidylinositol; and a phosphatidylcholine phospholipase C (PC-PLC), which is a lecithinase with enzymatic activity over phosphatidylcholine, phosphatidylserine, and phosphatidylethanolamine. PC-PLC is expressed as a protoenzyme and zinc-dependent metalloprotease Mpl is required for its maturation; so once free in the cytosol, the bacterium acquires the necessary nutrients for intracellular multiplication. Some studies have shown that LLO is a critical invasion factor, which perforates the plasma membrane of the host cell to activate the internalization of the bacterium in human hepatocytes. Moreover, other studies have shown that LLO fails to mediate the intracellular survival of L. monocytogenes in neutrophils, where early degranulation leads to the release of proteases such as matrix metalloproteinase (MMP)-8, degrading LLO and avoiding the perforation of the membranes.

L. monocytogenes causes a severe infection known as listeriosis, which is usually acquired after the intake of food contaminated with the microorganism. The disease mainly affects pregnant women, newborns, the elderly, and immunocompromised people, so it is rare for the disease to occur outside the aforementioned groups. Listeriosis is a mild disease in pregnant women, but it is severe in fetus and newborns. People over 65 years of age or immunosuppressed people can develop infection in the bloodstream (sepsis) or in the brain (meningitis or encephalitis). Sometimes the infection can affect bones, joints, thorax, and abdomen. Listeriosis can cause fever and diarrhea similar to that caused by other foodborne microorganisms and is rarely diagnosed. Pregnant women with listeriosis have fever, fatigue, and muscle pain (flu-like symptoms). During pregnancy, the organism can cause miscarriage, stillbirth, premature labor, and infection in the newborn. In the other risk groups, the symptoms are headaches, neck stiffness, confusion, loss of balance, seizures, fever and muscle pain. People with invasive listeriosis usually develop symptoms from 1 to 4 weeks after ingesting food contaminated with the bacterium; although symptoms have been reported after 70 days of exposure or on the same day of the poisoning. The disease is usually diagnosed by culturing the bacterium from tissues or fluids such as blood, cerebrospinal fluid, or placenta. From food products, this microorganism can be detected by various methods such as the use of chromogenic media; immunological methods, although some are nonspecific; molecular methods (hybridization, PCR, and real-time PCR); microarrays or biosensors; and also specific commercial methods. The detection of the plcA virulence gene coding for PI-PLC is generally employed to differentiate hemolytic and nonhemolytic strains. Pathogenic and nonpathogenic Listeria species can be differentiated by their activities of hemolysin or PI-PLC.

L. monocytogenes is a microorganism that can be present in many food products, mainly in dairy products, soft cheeses, cheeses made with unpasteurized milk, celery, cabbage, ice cream, hot dogs, and processed meats.

Infection with L. monocytogenes can be treated with antibiotics such as ampicillin, although penicillin is more effective. Some experts recommend the use of gentamicin in people with impaired immunity, including neonates, and in cases of meningitis and endocarditis. Ampicillin is only used in pregnant women with isolated listerial bacteremia. Other antibiotics that can be used are trimethoprim-sulfamethoxazole and vancomycin. Cephalosporins should not be used to treat listeriosis because they are ineffective against the microorganism.

The general guidelines to prevent listeriosis are similar to those recommended for other foodborne pathogens. For people at high risk, it is recommended not to consume soft cheeses such as Feta, Brie and Camembert, blue cheeses, or Mexican style cheeses (white cheese, fresh cheese, or panela cheese) unless they are made with pasteurized milk; it is also recommended not to consume smoked seafood, pâté or refrigerated meat spreads, hot dogs, processed meats or cold cuts, unless they have been reheated at high temperatures; these are just some of the food products that people at high risk should avoid.

Strategies for Disease Prevention

Multiple factors associated with the procurement, handling, and food preparation contribute to an increase in the likelihood of contamination, and consequently, consumer's poisoning. Due to the importance of foodborne diseases, the number of cases presented and their severity, it is necessary to know those measures that help preventing or avoiding them; or getting a disease caused by food poisoning related to bacterial toxins.

Toxigenic microorganisms arrive to food products by cross-contamination; they come from the environment or they belong to the normal microbiota, in the case of animals. Once the contaminated food is ingested and reaches the intestines, the microorganisms get established, colonize, and, if the strain is toxigenic, produce the toxins responsible for the damage. Likewise, an incubation process must occur prior to the first symptoms. To prevent the occurrence of such diseases, health care measures, especially hand hygiene of food handlers, should be carried out; in that way, all food sectors such as restaurants, manufacturing, and distribution companies, pay special attention to hygiene measures for food handling to prevent food handlers from inoculating the bacteria they carry on the skin on their hands. Along with other measures, they must ensure food safety, and for this, food sectors will establish policies and activities to ensure maximum quality and food safety throughout the food chain (from procurement and production to consumption).

Some of these standards are described and taken care by the Codex Alimentarius, which, together with the World Health Organization and the Food and Agriculture Organization of the United Nations, has the responsibility to develop and standardize the international food standards. Their objective is to ensure the quality of food products and to protect human health, as well as the correct and fair implementation of these standards. The standards of the Codex Alimentarius apply to processed, semiprocessed, or raw food products. In addition to all the factors used in food processing, food quality standards seek to ensure that food products are produced in hygienic conditions, and that they preserve their nutritional quality. The main standards include microbiological processes, regarding the use of food additives, pesticide use and pest control, as well as, the permissible limits of drugs or hormones used in animal production.

For proper handling of food products, facilities, materials, instruments, and equipment must be kept accessible for the cleaning and disinfection process, in order to prevent food contamination by toxigenic bacteria. Cleaning procedures will include the effective removal of food residues or other contaminants; these procedures must be continuous, because some microorganisms have the ability to settle on these surfaces and to survive in adverse conditions by forming biofilm, thus, cleaning with soap and water is not enough. The methods can be chemical, with alkaline and acidic detergents; and physical, with heat, turbulent washes, or vacuum washes. Moreover, brushes or sponges can be used to remove dirt; however, the correct method of use must be considered to ensure efficiency, as well as, not using the same cleaning instrument in areas of processed and unprocessed food. Detergents or disinfectant substances should be used under the conditions proposed by the manufacturer regarding the concentration and time of action, which will depend on the type of surface and the product's presentation (liquid, solid, or semisolid). Such cleaning processes will be subject to regular monitoring and quality control, registering the areas that were cleaned and the person responsible for the cleaning. The cleaning method will be used depending on what is intended to be cleaned; in the case of smooth surfaces, the use of disinfectant and sponges or brushes to remove residues will be enough; this is done in situ, contrary to those dismantled equipment that require to be cleaned piece by piece. All of the above related to the establishment's cleaning must be submitted in writing to the personnel responsible for this task for the correct and efficient implementation of cleaning methods.

Another important aspect in this sector is pest control. A variety of pests lurk at sites where food is produced; special care must be taken because in most cases these pests act as vehicles for toxigenic bacteria and other pathogens, endangering the consumer's health. The most common pests are rodents, flies, and cockroaches. To prevent the presence of pests, food facilities should avoid air vents and cracks; regarding food products, these should be stored in high places, inside sealed containers or bags to prevent rodents from smelling the food. For pest control, insect monitoring should be carried out on a continuous basis, through catch patches that may contain pheromones to attract insects, electric lamps against flying insects, among others. Of all insects, flies are the most common pest in food establishments, and they are an important source of disease transmission to food and other forms of food poisoning. It is important that food establishments eradicate flies pest to avoid any contamination of food products, in restaurants, kitchens, and other establishments where food is prepared; adhesive traps can be employed. Traps are used when managing rodent pests; however, an exhaustive planning must be done to determine the number of traps to be placed, as well as location; pest prevention include specifics such as covering air vents, avoiding cracks, and storage of food in high places, inside sealed containers or in bags to prevent rodents from smelling the food. At this point, the cleaning of the workplaces is of high importance, mainly the kitchen and the surfaces that are in contact with food, to ensure quality and food safety.

Food safety is a human right and an obligation of all the governments to ensure it; it refers to the preserved quality of food products without organoleptic alterations, the presence of chemical, physical, or biological pathogens, or other undesirable alterations in the products that may affect the consumer's health. In order to ensure this characteristic, good practices must be put into operation; identification and control of the potential sources of contamination by the establishment, proper storage of food by separating raw food from processed food, and handling of food products depending on their origin (animal or vegetable). Proper waste management and drainage installation need to be taken into account. Regarding the design and equipment distribution, and the areas where the food is prepared, raw food should be separated, and previously processed food should not be exposed in the same surface. Staff restrooms must be distant from food preparation areas to avoid fecal contamination. The use of suitable uniforms and footwear, air quality, ventilation, and temperature control are essential for a working environment that allows a good development of food processing, and reduces, as much as possible, food poisoning by toxigenic bacteria.

The Hazard Analysis and Critical Control Point (HACCP) system can be an efficient and systematic alternative to prevent toxico-infection; its function is to identify specific hazards and develop control measures to solve them, guaranteeing food safety by seven basic principles: identifying hazards and preventive measures, identifying critical control points, establishing limits, monitoring critical control points, using corrective measures, verifying processes, and registering the applied processes.

As a preventive measure to avoid food contamination and foodborne diseases, World Health Organization (WHO) proposes the five keys for food safety:

- Keep clean: It refers to washing hands before and during food preparation; after going to the toilet; washing and sanitizing surfaces and equipment for food preparation, and to keep them away from insects and animals.
- Separate raw and cooked food: Prepare in different surfaces raw and cooked food and use different equipment for each type of food.
- Cook thoroughly: Food cooked thoroughly allow the removal of bacteria and other pathogens; toxins produced by bacteria and pathogens can also be destroyed.
- Keep food at safe temperatures: Do not leave cooked food at room temperature for more than 2 h to avoid bacteria proliferation, and try not to store frozen food for long periods of time.
- Use safe water and raw materials: Safe treated water must be used when preparing food; use fresh food products and wash adequately. Pre-processed products such as pasteurized milk, should be used as directed and not be used beyond their expiry dates.

Mushroom Poisoning

Mushroom poisoning refers to harmful effects from ingestion of toxic substances present in a mushroom. These symptoms can vary from slight gastrointestinal discomfort to death in about 10 days. The toxins present are secondary metabolites produced by the fungus. Mushroom poisoning is usually the result of ingestion of wild mushrooms after misidentification of a toxic mushroom as an edible species. The most common reason for this misidentification is close resemblance in terms of colour and general morphology of the toxic mushrooms species with edible species. To prevent mushroom poisoning, mushroom gatherers familiarize themselves with the mushrooms they intend to collect, as well as with any similar-looking toxic species. The safety of eating wild mushrooms may depend on methods of preparation for cooking.

Amanita phalloides accounts for the majority of fatal mushroom poisonings worldwide.

Signs and Symptoms

Poisonous mushrooms contain a variety of different toxins that can differ markedly in toxicity. Symptoms of mushroom poisoning may vary from gastric upset to organ failure resulting in death. Serious symptoms do not always occur immediately after eating, often not until the toxin attacks the kidney or liver, sometimes days or weeks later.

The most common consequence of mushroom poisoning is simply gastrointestinal upset. Most "poisonous" mushrooms contain gastrointestinal irritants that cause vomiting and diarrhea (sometimes requiring hospitalization), but usually no long-term damage. However, there are a number of recognized mushroom toxins with specific, and sometimes deadly, effects.

Toxin	Toxicity	Effects
Alpha-amanitin	Deadly	Causes often fatal liver damage 1–3 days after ingestion. Principal toxin in the death cap.
Phallotoxin	Non-lethal	Causes extreme gastrointestinal upset. Found in various mushrooms.
Orellanine	Deadly	Redox cycler similar to paraquat. Causes kidney failure within 3 weeks after ingestion. Principal toxin in genus Cortinarius.
Muscarine	Potentially deadly	Causes SLUDGE syndrome. Found in various mushrooms. Antidote is atropine.
Monomethylhydrazine (MMH)	Deadly	Causes brain damage, seizures, gastrointestinal upset, and hemolysis. Metabolic poison. Principal toxin in genus Gyromitra. Antidote is large doses of intravenous pyridoxine hydrochloride.

Coprine	Non-lethal	Causes illness when consumed with alcohol. Principal toxin in genus Coprinus.
Ibotenic acid	Potentially deadly	Excitotoxin. Principal toxin in Amanita muscaria, A. pantherina, and A. gemmata.
Muscimol	Psychoactive	Causes CNS depression and hallucinations. Principal toxin in Amanita muscaria, A. pantherina, and A. gemmata.
Psilocybin and psilocin	Psychoactive	Causes CNS arousal and hallucinations. Principal effects in psilocybin mushrooms, many of which belonging to the genus Psilocybe (often used recreationally).
Arabitol	Non-lethal	Causes diarrhea in some people.
Bolesatine	Non-lethal	Causes gastrointestinal irritation, vomiting, nausea.
Ergotamine	Deadly	Affects the vascular system and can lead to loss of limbs and cardiac arrest. Found in genus Claviceps.

The period of time between ingestion and the onset of symptoms varies dramatically between toxins, some taking days to show symptoms identifiable as mushroom poisoning.

- Alpha-amanitin: For 6–12 hours, there are no symptoms. This is followed by a period of gastrointestinal upset (vomiting and profuse, watery diarrhea). This stage is caused primarily by the phallotoxins and typically lasts 24 hours. At the end of this second stage is when severe liver damage begins. The damage may continue for another 2–3 days. Kidney damage can also occur. Some patients will require a liver transplant. Amatoxins are found in some mushrooms in the genus *Amanita*, but are also found in some species of *Galerina* and *Lepiota*. Overall, mortality is between 10 and 15 percent. Recently, *Silybum marianum* or blessed milk thistle has been shown to protect the liver from amanita toxins and promote regrowth of damaged cells.
- Orellanine: This toxin causes no symptoms for 3–20 days after ingestion. Typically around day 11, the process of kidney failure begins, and is usually symptomatic by day 20. These symptoms can include pain in the area of the kidneys, thirst, vomiting, headache, and fatigue. A few species in the very large genus *Cortinarius* contain this toxin. People having eaten mushrooms containing orellanine may experience early symptoms as well, because the mushrooms often contain other toxins in addition to orellanine. A related toxin that causes similar symptoms but within 3–6 days has been isolated from *Amanita smithiana* and some other related toxic *Amanitas*.
- Muscarine: Muscarine stimulates the muscarinic receptors of the nerves and muscles. Symptoms include sweating, salivation, tears, blurred vision, palpitations, and, in high doses, respiratory failure. Muscarine is found in mushrooms of the genus *Omphalotus*, notably the jack o' Lantern mushrooms. It is also found in *A. muscaria*, although it is now known that the main effect of this mushroom is caused by ibotenic acid. Muscarine can also be found in some *Inocybe* species and *Clitocybe* species, in particular *Clitocybe dealbata*, and some red-pored *Boletes*.

- Gyromitrin: Stomach acids convert gyromitrin to monomethylhydrazine (MMH), a compound employed in rocket fuel. It affects multiple body systems. It blocks the important neurotransmitter GABA, leading to stupor, delirium, muscle cramps, loss of coordination, tremors, and/or seizures. It causes severe gastrointestinal irritation, leading to vomiting and diarrhea. In some cases, liver failure has been reported. It can also cause red blood cells to break down, leading to jaundice, kidney failure, and signs of anemia. It is found in mushrooms of the genus *Gyromitra*. A gyromitrin-like compound has also been identified in mushrooms of the genus *Verpa*.
- Coprine: Coprine is metabolized to a chemical that resembles disulfiram. It inhibits aldehyde dehydrogenase (ALDH), which, in general, causes no harm, unless the person has alcohol in their bloodstream while ALDH is inhibited. This can happen if alcohol is ingested shortly before or up to a few days after eating the mushrooms. In that case the alcohol cannot be completely metabolized, and the person will experience flushed skin, vomiting, headache, dizziness, weakness, apprehension, confusion, palpitations, and sometimes trouble breathing. Coprine is found mainly in mushrooms of the genus *Coprinus*, although similar effects have been noted after ingestion of *Clitocybe clavipes*.
- Ibotenic acid: Decarboxylates into muscimol upon ingestion. The effects of muscimol vary, but nausea and vomiting are common. Confusion, euphoria, or sleepiness are possible. Loss of muscular coordination, sweating, and chills are likely. Some people experience visual distortions, a feeling of strength, or delusions. Symptoms normally appear after 30 minutes to 2 hours and last for several hours. *A. muscaria*, the "Alice in Wonderland" mushroom, is known for the hallucinatory experiences caused by muscimol, but *A. pantherina* and *A. gemmata* also contain the same compound. While normally self-limiting, fatalities have been associated with *A. pantherina*, and consumption of a large number of any of these mushrooms is likely to be dangerous.
- Psilocybin: Dephosphorylates into the psychoactive psilocin upon ingestion, which acts as a psychedelic drug. Symptoms begin shortly after ingestion. The effects can include euphoria, visual and religious hallucinations, and heightened perception. However, some persons experience fear, agitation, confusion, and schizophrenia-like symptoms. All symptoms generally pass after several hours. Some (though not all) members of the genus *Psilocybe* contain psilocybin, as do some *Panaeolus*, *Copelandia*, *Conocybe*, *Gymnopilus*, and others. Some of these mushrooms also contain baeocystin, which has effects similar to psilocin.
- Arabitol: A sugar alcohol, similar to mannitol, which causes no harm in most people but causes gastrointestinal irritation in some. It is found in small amounts in oyster mushrooms, and considerable amounts in *Suillus* species and *Hygrophoropsis aurantiaca* (the "false chanterelle").

Causes

Amanita spp., immature, possibly poisonous, Amanita mushrooms.

New species of fungi are continuing to be discovered, with an estimated number of 800 new species registered annually. This, added to the fact that many investigations have recently reclassified some species of mushrooms from edible to poisonous has made older classifications insufficient at describing what now is known about the different species of fungi that are harmful to humans. Thus, contrary to what older registers state, it is now thought that of the approximately 100,000 known fungi species found worldwide, about 100 of them are poisonous to humans. However, by far the majority of mushroom poisonings are not fatal, and the majority of fatal poisonings are attributable to the *Amanita phalloides* mushroom.

Edible shaggy mane Coprinus comatus mushrooms.

A majority of these cases are due to mistaken identity. This is a common occurrence with *A. phalloides* in particular, due to its resemblance to the Asian paddy-straw mushroom, *Volvariella volvacea*. Both are light-colored and covered with a universal veil when young.

Amanitas can be mistaken for other species, as well, in particular when immature. On at least one occasion they have been mistaken for *Coprinus comatus*. In this case, the victim had some limited experience in identifying mushrooms, but did not take the time to correctly identify these particular mushrooms until after he began to experience symptoms of mushroom poisoning.

Amanitas, two examples of immature Amanitas, one deadly and one edible.

Researchers cautions puffball-hunters to beware of *Amanita* "eggs", which are *Amanitas* still entirely encased in their universal veil. *Amanitas* at this stage are difficult to distinguish from puffballs. Foragers are encouraged to always cut the fruiting bodies of suspected puffballs in half, as this will reveal the outline of a developing *Amanita* should it be present within the structure.

Puffball, an edible puffball mushroom, which closely resembles the immature Amanitas.

A majority of mushroom poisonings in general are the result of small children, especially toddlers in the "grazing" stage, ingesting mushrooms found in the lawn. While this can happen with any mushroom, *Chlorophyllum molybdites* is often implicated due to its preference for growing in lawns. *C. molybdites* causes severe gastrointestinal upset but is not considered deadly poisonous.

A few poisonings are the result of misidentification while attempting to collect hallucinogenic mushrooms for recreational use. In 1981, one fatality and two hospitalizations occurred following consumption of *Galerina autumnalis*, mistaken for a *Psilocybe* species. *Galerina* and *Psilocybe* species are both small, brown, and sticky, and can be found growing together. However, *Galerina* contains amatoxins, the same poison found in the deadly *Amanita* species. Another case reports kidney failure following ingestion of *Cortinarius orellanus*, a mushroom containing orellanine.

Jack-O-Lantern, a poisonous mushroom sometimes mistaken for a chanterelle.

It is natural that accidental ingestion of hallucinogenic species also occurs, but is rarely harmful when ingested in small quantities. Cases of serious toxicity have been reported in small children. *Amanita pantherina*, while containing the same hallucinogens as *Amanita muscaria* (e.g., ibotenic acid and muscimol), has been more commonly associated with severe gastrointestinal upset than its better-known counterpart.

Chanterelle, edible.

Although usually not fatal, *Omphalotus* spp., "Jack-o-lantern mushrooms," are another cause of sometimes significant toxicity. They are sometimes mistaken for chanterelles. Both are bright-orange and fruit at the same time of year, although *Omphalotus* grows on wood and has true gills rather than the veins of a *Cantharellus*. They contain toxins known as illudins, which causes gastrointestinal symptoms.

Bioluminescent species are generally inedible and often mildly toxic. *Clitocybe dealbata*, which is occasionally mistaken for an oyster mushroom or other edible species contains muscarine.

Toxicities can also occur with collection of morels. Even true morels, if eaten raw, will cause gastrointestinal upset. Typically, morels are thoroughly cooked before eating. *Verpa bohemica*, although referred to as "thimble morels" or "early morels" by some, have caused toxic effects in some individuals. *Gyromitra* spp., "false morels", are deadly poisonous if eaten raw. They contain a toxin called gyromitrin, which can cause neurotoxicity, gastrointestinal toxicity, and destruction of the blood cells. The Finns consume *Gyromitra esculenta* after parboiling, but this may not render the mushroom entirely safe, resulting in its being called the "fugu of the Finnish cuisine".

A more unusual toxin is coprine, a disulfiram-like compound that is harmless unless ingested within a few days of ingesting alcohol. It inhibits aldehyde dehydrogenase, an enzyme required for breaking down alcohol. Thus, the symptoms of toxicity are similar to being hung over—flushing, headache, nausea, palpitations, and, in severe cases, trouble breathing. *Coprinus* species, including *Coprinopsis atramentaria*, contain coprine. *Coprinus comatus* does not, but it is best to avoid mixing alcohol with other members of this genus.

Recently, poisonings have also been associated with *Amanita smithiana*. These poisonings may be due to orellanine, but the onset of symptoms occurs in 4 to 11 hours, which is much quicker than the 3 to 20 days normally associated with orellanine.

Paxillus involutus is also inedible when raw, but is eaten in Europe after pickling or parboiling. However, after the death of the German mycologist Dr Julius Schäffer, it was discovered that the mushroom contains a toxin that can stimulate the immune system to attack its own red blood cells. This reaction is rare, but can occur even after safely eating the mushroom for many years. Similarly, *Tricholoma equestre* was widely considered edible and good, until it was connected with rare cases of rhabdomyolysis.

In the fall of 2004, thirteen deaths were associated with consumption of *Pleurocybella porrigens* or "angel's wings". In general, these mushrooms are considered edible. All the victims died of an acute brain disorder, and all had pre-existing kidney disease. The exact cause of the toxicity was not known at this time and the deaths cannot be definitively attributed to mushroom consumption.

However, mushroom poisoning is not always due to mistaken identity. For example, the highly toxic ergot *Claviceps purpurea*, which grows on rye, is sometimes ground up with rye, unnoticed, and later consumed. This can cause devastating, even fatal effects, which is called ergotism.

Cases of idiosyncratic or unusual reactions to fungi can also occur. Some are probably due to allergy, others to some other kind of sensitivity. It is not uncommon for an individual person to experience gastrointestinal upset associated with one particular mushroom species or genus.

Some mushrooms might concentrate toxins from their growth substrate, such as Chicken of the Woods growing on yew trees.

Poisonous Mushrooms

Of the most lethal mushrooms, three—the death cap (*A. phalloides*), destroying angels (*A. virosa* and *A. bisporigera*), and the fool's mushroom (*A. verna*)—belong to the genus *Amanita*, and two more—the deadly webcap (*C. rubellus*), and the fool's webcap

(*C. orellanus*)—are from the genus *Cortinarius*. Several species of Galerina, Lepiota, and Conocybe also contain lethal amounts of amatoxins. Deadly species are listed in the List of deadly fungi.

The following species may cause great discomfort, sometimes requiring hospitalization, but are not considered deadly.

- Amanita muscaria (fly agaric): Contains the psychoactive muscimol and the neurotoxin ibotenic acid. Ibotenic acid decarboxylates into muscimol upon curing of the mushroom, rendering it relatively non-toxic, though death via respiratory depression is possible. Muscimol intoxication is often considered unpleasant and undesirable, however, and as such has seen little recreational use compared to the unrelated psilocybin mushroom, though it has been used as an entheogen by the native people of Siberia.
- Amanita pantherina (panther mushroom): Contains similar toxins as A. muscaria, but is associated with more fatalities than A. muscaria.
- Chlorophyllum molybdites (greengills): Causes intense gastrointestinal upset.
- Entoloma (pinkgills): Some species are highly poisonous, such as livid entoloma (Entoloma sinuatum), Entoloma rhodopolium, and Entoloma nidorosum. Symptoms of intense gastrointestinal upset appear after 20 minutes to 4 hours, caused by an unidentified gastrointestinal irritant.
- Many Inocybe species such as Inocybe fastigiata and Inocybe geophylla contain muscarine, while Inocybe erubescens is the only one known to have caused death.
- Some white Clitocybe species, including C. rivulosa and C. dealbata: Contain muscarine.
- Tricholoma pardinum, Tricholoma tigrinum (tiger tricholoma): Gastrointestinal upset due to an unidentified toxin, begins in 15 minutes to 2 hours and lasts 4 to 6 days.
- Tricholoma equestre (man-on-horseback): Until recently thought edible and good, can lead to rhabdomyolysis after repeated consumption.
- Hypholoma fasciculare/Naematoloma fasciculare (sulfur tuft): Usually causes gastrointestinal upset, but the toxins fasciculol E and F could lead to paralysis and death.
- Paxillus involutus (brown roll-rim): Once thought edible, but now found to destroy red blood cells with regular or long-term consumption.
- Rubroboletus satanas (Devil's bolete), Suillellus luridus, Rubroboletus legaliae, Chalciporus piperatus, Neoboletus luridiformis, Rubroboletus pulcherrimus: Gastrointestinal irritation. Of these, only R. pulcherrimus has been implicated in a death. Many books list N. luridiformis as edible, but Arora lists it as "to be avoided".

- Hebeloma crustuliniforme (known as poison pie or fairy cakes): Causes gastrointestinal symptoms such as nausea and vomiting.
- Russula emetica (the sickener): As its name implies, causes rapid vomiting. Other Russulas with a peppery taste (Russula silvicola, Russula mairei) will likely do the same.
- Agaricus hondensis, Agaricus californicus, Agaricus praeclaresquamosus, Agaricus xanthodermus: Cause vomiting and diarrhea in most people, although some people seem to be immune.
- Lactifluus piperatus, Lactarius torminosus, Lactarius rufus: These and other peppery-tasting milk-caps are pickled and eaten in Scandinavia, but are indigestible or poisonous unless correctly prepared.
- Lactarius vinaceorufescens, Lactarius uvidus: Reported to be poisonous. Arora reports that all yellow- or purple-staining Lactarius are "best avoided".
- Ramaria gelatinosa: Causes indigestion in many people, although some seem immune.
- Gomphus floccosus (the scaly chanterelle): Causes gastric upset in many people, although some eat it without problems. *G. floccosus* is sometimes confused with the chanterelle.

Prognosis and Treatment

Some mushrooms contain less toxic compounds and, therefore, are not severely poisonous. Poisonings by these mushrooms may respond well to treatment. However, certain types of mushrooms, contain very potent toxins and are very poisonous; so even if symptoms are treated promptly mortality is high. With some toxins, death can occur in a week or a few days. Although a liver or kidney transplant may save some patients with complete organ failure, in many cases there are no organs available. Patients hospitalized and given aggressive support therapy almost immediately after ingestion of amanitin-containing mushrooms have a mortality rate of only 10%, whereas those admitted 60 or more hours after ingestion have a 50–90% mortality rate.

Scombroid Food Poisoning

Scombroid food poisoning, also known as simple scombroid, is a foodborne illness that typically results from eating spoiled fish. Symptoms may include flushed skin, headache, itchiness, blurred vision, abdominal cramps, and diarrhea. Onset of symptoms is typically 10 to 60 minutes after eating and can last for up to two days. Rarely, breathing problems or an irregular heartbeat may occur.

Scombroid occurs from eating fish high in histamine due to inappropriate storage or processing. Fish commonly implicated include tuna, mackerel, mahi mahi, sardine, anchovy, herring, bluefish, amberjack, and marlin. These fish naturally have high levels of histidine, which is converted to histamine when bacterial growth occurs during improper storage. Subsequent cooking, smoking, or freezing does not eliminate the histamine. Diagnosis is typically based on the symptoms and may be supported by a normal blood tryptase. If a number of people who eat the same fish develop symptoms, the diagnosis is more likely.

Prevention is by refrigerating or freezing fish right after it is caught. Treatment is generally with antihistamines such as diphenhydramine and ranitidine. Epinephrine may be used for severe symptoms. Along with ciguatera fish poisoning, it is one of the most common type of seafood poisoning. It occurs globally in both temperate and tropical waters. Only one death has been reported. The condition was first described in 1799.

Signs and Symptoms

Symptoms typically occur within 10–30 minutes of ingesting the fish and generally are self-limited. People with asthma are more vulnerable to respiratory problems such as wheezing or bronchospasms. However, symptoms may show over two hours after eating a spoiled dish. They usually last for about 10 to 14 hours, and rarely exceed one to two days.

Initial

The first signs of poisoning suggest an allergic reaction with these symptoms:

- Facial flushing/sweating,
- Burning-peppery taste sensations in the mouth and throat,
- Dizziness,
- Nausea,
- Headache,
- Tachycardia,
- Cold-like symptoms.

Additional Symptoms

The above symptoms can advance to:

- Facial rash (intense itching may accompany the rash),
- Torso or body rash: The rash associated with scombroid poisoning is a form of urticaria, but most commonly does not include wheals (patchy areas of skin-swelling also known as hives) that may be seen in true allergies,

- Edema (generalized if it occurs at all),
- Short-term diarrhea,
- Abdominal cramps.

Severe

In the worst cases, the poisoning may cause:

- Blurred vision,
- Respiratory distress,
- Swelling of the tongue.

In rare cases, the poisoning may result in death.

Causes

Unlike many types of food poisoning, scombroid form is not brought about by ingestion of a pathogen. Histidine is an amino acid that exists naturally in many types of food, including fish. At temperatures above 16 °C (60 °F), histidine is converted to the biogenic amine histamine via the enzyme histidine decarboxylase produced by symbiotic bacteria such as *Morganella morganii* (this is one reason why fish should be stored in the freezer). Histamine is not destroyed by normal cooking temperatures, so even properly cooked fish can still result in poisoning. Histamine is the main natural chemical responsible for true allergic reactions, so the symptoms produced are almost identical to a food allergy. Rarely cheese may be involved.

Diagnosis

Differentiating scombroid from a fish allergy can be difficult, as both present with similar symptoms. In scombroid, blood tryptase is generally normal, while in an allergic reaction, it is elevated.

Prevention

Prevention is by refrigerating or freezing fish right after they are killed. Antihistamines may also be taken before eating seafood.

VIBRIO VULNIFICUS

Vibrio vulnificus food poisoning is caused by Vibrio vulnificus, a bacterium that lives in warm seawater. The condition is rare.

What causes Vibrio Vulnificus Food Poisoning?

Vibrio vulnificus food poisoning occurs when you eat seafood infected with the bacteria or you have an open wound that is exposed to them. The bacteria are frequently found in oysters and other shellfish in warm coastal waters during the summer months. People who have weak immune systems, especially those with long-term (chronic) liver disease, are at greater risk for this condition than other people.

What are the Symptoms?

In healthy people, Vibrio vulnificus food poisoning can cause vomiting, diarrhea, and abdominal (belly) pain. In people who have weak immune systems, the bacteria can infect the bloodstream, causing a severe and life-threatening illness. Symptoms include fever and chills, decreased blood pressure (septic shock), and blistering skin wounds. The infection is especially dangerous to people who have long-term (chronic) liver disease.

If an open wound is exposed to the bacteria (such as from warm seawater), sores may develop. People with weak immune systems are at risk for the bacteria moving into the bloodstream.

How is it Treated?

You treat Vibrio vulnificus food poisoning by managing complications until it passes. Dehydration caused by diarrhea and vomiting is the most common complication. In people who have weak immune systems, or in people who have severe symptoms, antibiotics may be used.

To prevent dehydration, take frequent sips of a rehydration drink (such as Pedialyte). Try to drink a cup of water or rehydration drink for each large, loose stool you have. Soda and fruit juices have too much sugar and not enough of the important electrolytes that are lost during diarrhea, and they should not be used to rehydrate.

Try to stay with your normal diet as much as possible. Eating your usual diet will help you to get enough nutrition. Doctors believe that eating a normal diet will also help you feel better faster. But try to avoid foods that are high in fat and sugar. Also avoid spicy foods, alcohol, and coffee for 2 days after all symptoms have disappeared.

The best way to prevent this type of food poisoning is to not eat raw oysters or other raw shellfish and to cook all shellfish (oysters, clams, mussels) thoroughly.

Boil shucked oysters for at least 3 minutes or fry them in oil for at least 10 minutes at 375 °F (191 °C). For shellfish in the shell, either:

- Boil until the shells open and continue boiling for 5 more minutes.
- Steam until the shells open and then continue cooking for 9 more minutes.

Do not eat those shellfish that do not open during cooking.

You should also,

- Avoid cross-contamination of cooked seafood and other foods with raw seafood and juices from raw seafood. Don't prepare them in the same place. And don't use the same cutting board when preparing them.
- Eat shellfish immediately after cooking, and refrigerate leftovers.
- Avoid exposing open wounds or broken skin to warm saltwater or brackish water or to raw shellfish harvested from such waters.
- Wear protective clothing, such as gloves, when you handle raw shellfish.

FOOD ALLERGY

A food allergy is an abnormal immune response to food. The symptoms of the allergic reaction may range from mild to severe. They may include itchiness, swelling of the tongue, vomiting, diarrhea, hives, trouble breathing, or low blood pressure. This typically occurs within minutes to several hours of exposure. When the symptoms are severe, it is known as anaphylaxis. A food intolerance and food poisoning are separate conditions, not due to an immune response.

Common foods involved include cow's milk, peanuts, eggs, shellfish, fish, tree nuts, soy, wheat, rice, and fruit. The common allergies vary depending on the country. Risk factors include a family history of allergies, vitamin D deficiency, obesity, and high levels of cleanliness. Allergies occur when immunoglobulin E (IgE), part of the body's immune system, binds to food molecules. A protein in the food is usually the problem. This triggers the release of inflammatory chemicals such as histamine. Diagnosis is usually based on a medical history, elimination diet, skin prick test, blood tests for food-specific IgE antibodies, or oral food challenge.

Early exposure to potential allergens may be protective. Management primarily involves avoiding the food in question and having a plan if exposure occurs. This plan may include giving adrenaline (epinephrine) and wearing medical alert jewelry. The benefits of allergen immunotherapy for food allergies is unclear, thus is not recommended as of 2015. Some types of food allergies among children resolve with age, including that to milk, eggs, and soy; while others such as to nuts and shellfish typically do not.

In the developed world, about 4% to 8% of people have at least one food allergy. They are more common in children than adults and appear to be increasing in frequency. Male children appear to be more commonly affected than females. Some allergies more commonly develop early in life, while others typically develop in later life. In developed countries, a large proportion of people believe they have food allergies when they actually do not have them. The declaration of the presence of trace amounts of allergens in foods is mandatory only in Brazil.

Signs and Symptoms

Food allergies usually have a fast onset (from seconds to one hour) and may include:

- Rash;
- Hives;
- Itching of mouth, lips, tongue, throat, eyes, skin, or other areas;
- Swelling (angioedema) of lips, tongue, eyelids, or the whole face;
- Difficulty swallowing;
- Runny or congested nose;
- Hoarse voice;
- Wheezing and/or shortness of breath;
- Diarrhea, abdominal pain, and/or stomach cramps;
- Lightheadedness;
- Fainting;
- Nausea;
- Vomiting.

In some cases, however, onset of symptoms may be delayed for hours.

Symptoms can vary. The amount of food needed to trigger a reaction also varies. Serious danger regarding allergies can begin when the respiratory tract or blood circulation is affected. The former can be indicated through wheezing and cyanosis. Poor blood circulation leads to a weak pulse, pale skin and fainting.

A severe case of an allergic reaction, caused by symptoms affecting the respiratory tract and blood circulation, is called anaphylaxis. When symptoms are related to a drop in blood pressure, the person is said to be in anaphylactic shock. Anaphylaxis occurs when IgE antibodies are involved, and areas of the body that are not in direct contact with the food become affected and show symptoms. Those with asthma or an allergy to peanuts, tree nuts, or seafood are at greater risk for anaphylaxis.

Cause

Although sensitivity levels vary by country, the most common food allergies are allergies to milk, eggs, peanuts, tree nuts, seafood, shellfish, soy, and wheat. These are often referred to as "the big eight". Allergies to seeds – especially sesame – seem to be increasing in many countries. An example an allergy more common to a particular region is that to rice in East Asia where it forms a large part of the diet.

One of the most common food allergies is a sensitivity to peanuts, a member of the bean family. Peanut allergies may be severe, but children with peanut allergies sometimes outgrow them. Tree nuts, including almonds, brazil nuts, cashews, coconuts, hazelnuts, macadamia nuts, pecans, pistachios, pine nuts, and walnuts, are also common allergens. Sufferers may be sensitive to one particular tree nut or to many different ones. Also, seeds, including sesame seeds and poppy seeds, contain oils where protein is present, which may elicit an allergic reaction.

Egg allergies affect about one in 50 children but are frequently outgrown by children when they reach age five. Typically, the sensitivity is to proteins in the white, rather than the yolk.

Milk from cows, goats, or sheep is another common food allergen, and many sufferers are also unable to tolerate dairy products such as cheese. A small portion of children with a milk allergy, roughly 10%, have a reaction to beef. Beef contains a small amount of protein that is also present in cow's milk.

Seafood is one of the most common sources of food allergens; people may be allergic to proteins found in fish, crustaceans, or shellfish.

Other foods containing allergenic proteins include soy, wheat, fruits, vegetables, maize, spices, synthetic and natural colors, and chemical additives.

Balsam of Peru, which is in various foods, is in the "top five" allergens most commonly causing patch test reactions in people referred to dermatology clinics.

Sensitization

Sensitization can occur through the gastrointestinal tract, respiratory tract and possibly the skin. Damage to the skin in conditions such as eczema has been proposed as a risk factor for sensitization. An Institute of Medicine report says that food proteins contained in vaccines, such as gelatin, milk, or egg can cause sensitization (development of allergy) in vaccine recipients, to those food items.

Atopy

Food allergies develop more easily in people with the atopic syndrome, a very common combination of diseases: allergic rhinitis and conjunctivitis, eczema, and asthma. The syndrome has a strong inherited component; a family history of allergic diseases can be indicative of the atopic syndrome.

Cross-reactivity

Some children who are allergic to cow's milk protein also show a cross-sensitivity to soy-based products. Some infant formulas have their milk and soy proteins hydrolyzed, so when taken by infants, their immune systems do not recognize the allergen and they can safely consume the product. Hypoallergenic infant formulas can be based on proteins partially predigested to a less antigenic form. Other formulas, based on free amino acids, are the least antigenic and provide complete nutritional support in severe forms of milk allergy.

People with latex allergy often also develop allergies to bananas, kiwifruit, avocados, and some other foods.

Pathophysiology

Conditions caused by food allergies are classified into three groups according to the mechanism of the allergic response:

- IgE-mediated (classic): The most common type, occurs shortly after eating and may involve anaphylaxis.
- Non-IgE mediated: Characterized by an immune response not involving immunoglobulin E; may occur some hours after eating, complicating diagnosis.
- IgE and/or non-IgE-mediated: A hybrid of the above two types.

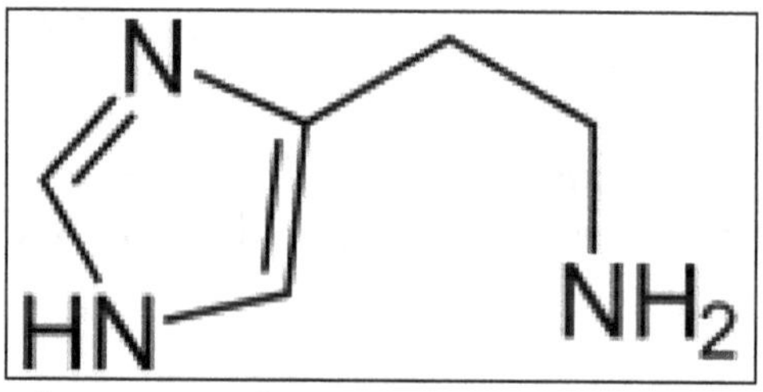

A histamine, the structure shown, causes a person to feel itchy during an allergic reaction. A common medication to stop this is an antihistamine, which fights the histamines in the person's system.

Allergic reactions are hyperactive responses of the immune system to generally innocuous substances. When immune cells encounter the allergenic protein, IgE antibodies are produced; this is similar to the immune system's reaction to foreign pathogens. The IgE antibodies identify the allergenic proteins as harmful and initiate the allergic reaction. The harmful proteins are those that do not break down due to the strong bonds of the protein. IgE antibodies bind to a receptor on the surface of the protein, creating a tag, just as a virus or parasite becomes tagged. Why some proteins do not denature and subsequently trigger allergic reactions and hypersensitivity while others do is not entirely clear.

Hypersensitivities are categorized according to the parts of the immune system that are attacked and the amount of time it takes for the response to occur. The four types of hypersensitivity reaction are: type 1, immediate IgE-mediated; type 2, cytotoxic; type 3, immune complex-mediated; and type 4, delayed cell-mediated. The pathophysiology of allergic responses can be divided into two phases. The first is an acute response that occurs immediately after exposure to an allergen. This phase can either subside or progress into a "late-phase reaction" which can substantially prolong the symptoms of a response, and result in tissue damage.

Many food allergies are caused by hypersensitivities to particular proteins in different foods. Proteins have unique properties that allow them to become allergens, such as stabilizing forces in their tertiary and quaternary structures which prevent degradation during digestion. Many theoretically allergenic proteins cannot survive the destructive environment of the digestive tract, thus do not trigger hypersensitive reactions.

Acute Response

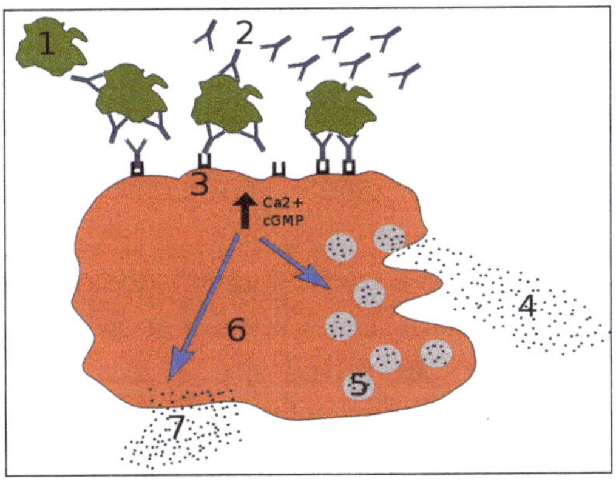

In figure, Degranulation process in allergy:

1. Antigen,
2. IgE antibody,
3. FcεRI receptor,
4. Preformed mediators (histamine, proteases, chemokines, heparin),
5. Granules,
6. Mast cell,
7. Newly formed mediators (prostaglandins, leukotrienes, thromboxanes, PAF).

In the early stages of allergy, a type I hypersensitivity reaction against an allergen, encountered for the first time, causes a response in a type of immune cell called a T_H2 lymphocyte, which belongs to a subset of T cells that produce a cytokine called interleukin-4 (IL-4). These T_H2 cells interact with other lymphocytes called B cells, whose role is the production of antibodies. Coupled with signals provided by IL-4, this interaction stimulates the B cell to begin production of a large amount of a particular type of antibody known as IgE. Secreted IgE circulates in the blood and binds to an IgE-specific receptor (a kind of Fc receptor called FcεRI) on the surface of other kinds of immune cells called mast cells and basophils, which are both involved in the acute inflammatory response. The IgE-coated cells, at this stage, are sensitized to the allergen.

If later exposure to the same allergen occurs, the allergen can bind to the IgE molecules held on the surface of the mast cells or basophils. Cross-linking of the IgE and Fc receptors occurs when more than one IgE-receptor complex interacts with the same allergenic molecule, and activates the sensitized cell. Activated mast cells and basophils undergo a process called degranulation, during which they release histamine and other inflammatory chemical mediators (cytokines, interleukins, leukotrienes, and prostaglandins) from their granules into the surrounding tissue causing several systemic effects, such as vasodilation, mucous secretion, nerve stimulation, and smooth-muscle contraction. This results in rhinorrhea, itchiness, dyspnea, and anaphylaxis. Depending on the individual, the allergen, and the mode of introduction, the symptoms can be system-wide (classical anaphylaxis), or localized to particular body systems.

Late-phase Response

After the chemical mediators of the acute response subside, late-phase responses can often occur due to the migration of other leukocytes such as neutrophils, lymphocytes, eosinophils, and macrophages to the initial site. The reaction is usually seen 2–24 hours after the original reaction. Cytokines from mast cells may also play a role in the persistence of long-term effects.

Diagnosis

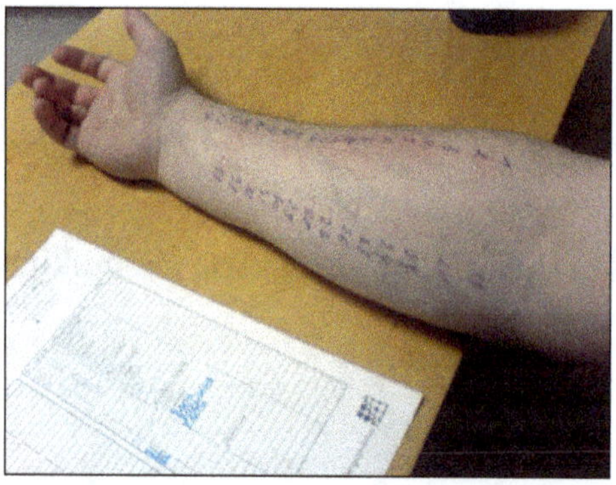

Skin testing on the arm is a common way for detecting an allergy, but it is not as effective as other tests.

Diagnosis is usually based on a medical history, elimination diet, skin prick test, blood tests for food-specific IgE antibodies, or oral food challenge.

- For skin-prick tests, a tiny board with protruding needles is used. The allergens are placed either on the board or directly on the skin. The board is then placed on the skin, to puncture the skin and for the allergens to enter the body. If a hive appears, the person is considered positive for the allergy. This test only works for IgE antibodies. Allergic reactions caused by other antibodies cannot be detected through skin-prick tests.

  Skin-prick testing is easy to do and results are available in minutes. Different allergists may use different devices for testing. Some use a "bifurcated needle", which looks like a fork with two prongs. Others use a "multitest", which may look like a small board with several pins sticking out of it. In these tests, a tiny amount of the suspected allergen is put onto the skin or into a testing device, and the device is placed on the skin to prick, or break through, the top layer of skin. This puts a small amount of the allergen under the skin. A hive will form at any spot where the person is allergic. This test generally yields a positive or negative result. It is good for quickly learning if a person is allergic to a particular food or not, because it detects IgE. Skin tests cannot predict if a reaction would occur or what kind of reaction might occur if a person ingests that particular allergen. They can, however, confirm an allergy in light of a patient's history of reactions to a particular food. Non-IgE-mediated allergies cannot be detected by this method.

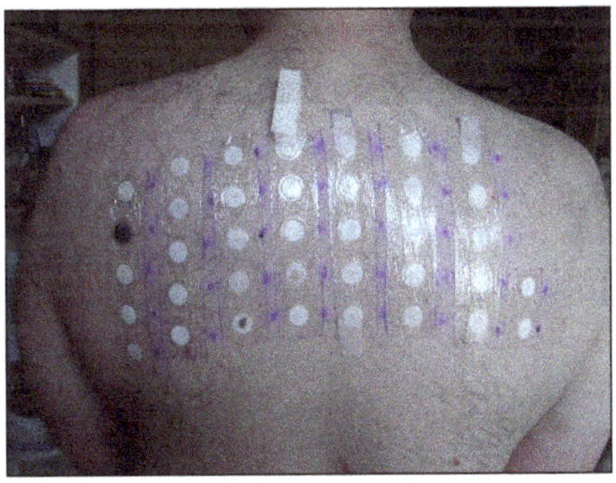

Patch test.

- Patch testing is used to determine if a specific substance causes allergic inflammation of the skin. It tests for delayed food reactions.

- Blood testing is another way to test for allergies; however, it poses the same disadvantage and only detects IgE allergens and does not work for every possible allergen. Radioallergosorbent testing (RAST) is used to detect IgE antibodies present to a certain allergen. The score taken from the RAST is compared to predictive values, taken from a specific type of RAST. If the score is higher than the predictive values, a great chance the allergy is present in the person exists. One advantage of this test is that it can test many allergens at one time.

A CAP-RAST has greater specificity than RAST; it can show the amount of IgE present to each allergen. Researchers have been able to determine "predictive values" for certain foods, which can be compared to the RAST results. If a person's RAST score is higher than the predictive value for that food, over a 95% chance exists that patients will have an allergic reaction (limited to rash and anaphylaxis reactions) if they ingest that food. Currently, predictive values are available for milk, egg, peanut, fish, soy, and wheat. Blood tests allow for hundreds of allergens to be screened from a single sample, and cover food allergies as well as inhalants. However, non-IgE-mediated allergies cannot be detected by this method. Other widely promoted tests such as the antigen leukocyte cellular antibody test and the food allergy profile are considered unproven methods, the use of which is not advised.

- Food challenges test for allergens other than those caused by IgE allergens. The allergen is given to the person in the form of a pill, so the person can ingest the allergen directly. The person is watched for signs and symptoms. The problem with food challenges is that they must be performed in the hospital under careful watch, due to the possibility of anaphylaxis.

Food challenges, especially double-blind, placebo-controlled food challenges, are the gold standard for diagnosis of food allergies, including most non-IgE-mediated reactions, but is rarely done. Blind food challenges involve packaging the suspected allergen into a capsule, giving it to the patient, and observing the patient for signs or symptoms of an allergic reaction.

The recommended method for diagnosing food allergy is to be assessed by an allergist. The allergist will review the patient's history and the symptoms or reactions that have been noted after food ingestion. If the allergist feels the symptoms or reactions are consistent with food allergy, he/she will perform allergy tests. Additional diagnostic tools for evaluation of eosinophilic or non-IgE mediated reactions include endoscopy, colonoscopy, and biopsy.

Differential Diagnosis

Important differential diagnoses are:

- Lactose intolerance generally develops later in life, but can present in young patients in severe cases. It is due to an enzyme deficiency (lactase) and not allergy, and occurs in many non-Western people.
- Celiac disease: While it is caused by a permanent intolerance to gluten (present in wheat, rye, barley and oats), is not an allergy nor simply an intolerance, but a chronic, multiple-organ autoimmune disorder primarily affecting the small intestine.
- Irritable bowel syndrome.

- C1 Esterase inhibitor deficiency (hereditary angioedema), a rare disease, generally causes attacks of angioedema, but can present solely with abdominal pain and occasional diarrhea.

Prevention

Breastfeeding for more than four months may prevent atopic dermatitis, cow's milk allergy, and wheezing in early childhood. Early exposure to potential allergens may be protective. Specifically, early exposure to eggs and peanuts reduces the risk of allergies to these. Guidelines suggest introducing peanuts as early as 4–6 months and include precautionary measures for high-risk infants. The former guidelines, advising delaying the introduction of peanuts, are now thought to have contributed to the increase in peanut allergy seen recently.

To avoid an allergic reaction, a strict diet can be followed. It is difficult to determine the amount of allergenic food required to elicit a reaction, so complete avoidance should be attempted. In some cases, hypersensitive reactions can be triggered by exposures to allergens through skin contact, inhalation, kissing, participation in sports, blood transfusions, cosmetics, and alcohol.

Inhalation Exposure

Allergic reactions to airborne particles or vapors of known food allergens have been reported as an occupational consequence of people working in the food industry, but can also take place in home situations, restaurants, or confined spaces such as airplanes. According to two reviews, respiratory symptoms are common, but in some cases there has been progression to anaphylaxis. The most frequent reported cases of reactions by inhalation of allergenic foods were due to peanut, seafood, legumes, tree nut, and cow's milk. Steam rising from cooking of lentils, green beans, chickpeas and fish has been well documented as triggering reactions, including anaphylactic reactions. One review mentioned case study examples of allergic responses to inhalation of other foods, including examples in which oral consumption of the food is tolerated.

DIARRHEA

Diarrhea is a common problem that can come on suddenly or be a chronic complaint. The possible causes of diarrhea include food poisoning, infections, food allergies or intolerances, and medication. Some conditions that cause chronic diarrhea run in families or, rarely, have a genetic basis. You can reduce your risk of acute diarrhea by learning good food safety and cleanliness habits, whether at home or while traveling.

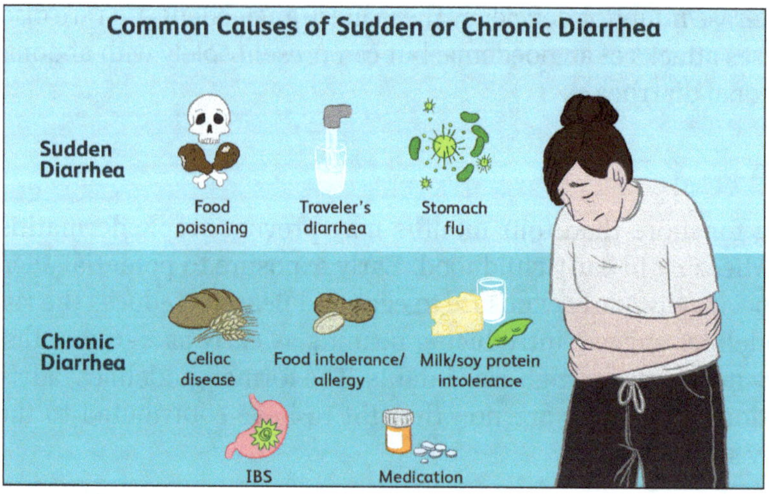

Common Causes

The most common cause of diarrhea, especially diarrhea that starts suddenly (acute diarrhea), is an infection—this can mean an infection from bacteria, a virus, or a parasite.[1] There are many possible causes of acute diarrhea, but the three most common are:

- Food poisoning,
- Traveler's diarrhea,
- Stomach flu.

You may also have acute diarrhea due to side effects of medications, including antibiotics, antacids that contain magnesium, and cancer treatment medications.

Children are especially at risk for diarrhea as they tend to put objects in their mouths and may not have developed good handwashing habits. They are at high risk of dehydration from diarrhea, so they should be seen by a doctor.

Food Poisoning

Food poisoning occurs when you eat food that is contaminated with bacteria. [2] The bacteria build up toxins in the food that make you sick. Causes of food poisoning are poor sanitation, improper food handling, or food being stored at the wrong temperature.

- How long diarrhea lasts: Usually less than two days.
- Triggered by: Toxins in food.
- Symptoms appear within: Two to six hours of ingesting the food.
- Appearance: Explosive, watery.
- Other symptoms: Abdominal cramps, fever, vomiting, weakness.

Food Poisoning Signs and Symptoms

Traveler's Diarrhea

Traveler's diarrhea is caused by eating food or drinking water that is contaminated with bacteria or parasites. Most traveler's diarrhea will get better with home care in a few days. If you have diarrhea and recently traveled to a less-developed country or drank untreated water from a stream, river, or pond (in the U.S. or elsewhere), call your doctor.

- How long diarrhea lasts: Usually less than one week.
- Triggered by: Food or water that is contaminated by bacteria, viruses, or parasites.
- Symptoms appear within: 12 to 24 hours.
- Appearance: Explosive, watery, sometimes contains mucus or blood.
- Other symptoms: Possibly vomiting and fever.

Stomach Flu

Stomach flu, also known as gastroenteritis, is caused by a virus, but not the same that causes seasonal flu (influenza). Examples of viruses that can cause stomach flu are rotavirus and norovirus. Gastroenteritis also can be caused by a bacterium or a parasite. You can generally care for your stomach flu with home treatment. Young children, the elderly, and people with compromised immune systems are at risk for dehydration[4], and should be watched closely for signs of more serious complications.

- How long diarrhea lasts: Usually three to eight days.
- Triggered by: A virus, bacterium, or parasite.
- Symptoms appear within: two days after exposure.
- Appearance: Watery.
- Other symptoms: Vomiting, fever, achiness.

Causes of Chronic Diarrhea

Diarrhea that goes on for weeks or months may be caused by an infection, or it may be caused by an underlying medical condition. Here are some of the possible causes of chronic diarrhea, and there are many more.

Celiac Disease

If you have untreated celiac disease, you may have a hard time linking your symptoms with a specific food because your gut is damaged and you may experience symptoms all the time.

- How long diarrhea lasts: More than four weeks.
- Triggered by: Gluten, but may hard to pinpoint your symptoms to a specific meal.
- Appearance: Large, bad-smelling stools that float in water and may appear greasy.
- Other symptoms: Unintended weight loss, lack of energy, lack of growth in children, plus many other possible symptoms.

Food Allergy

Symptoms of classic Ig-E mediated food allergy begin within minutes to hours of eating a trigger food. It is possible to be allergic to any food, but a few foods cause the most common food allergies.

- How long diarrhea lasts: Usually less than 24 hours.
- Triggered by: A specific food.
- Symptoms appear within: Two hours.
- Appearance: Watery, may contain blood.
- Other symptoms: Hives; vomiting; swelling of face, tongue or throat; eczema.

Food Intolerance

Food intolerance is caused by a lack of the enzymes needed to digest a specific food. Lactose intolerance, the inability to digest the sugars in milk, is the most common, but it's possible to be intolerant of other foods as well.

- How long diarrhea lasts: More than four weeks.
- Triggered by: A specific food.
- Symptoms appear within: two to 12 hours.
- Appearance: Watery, sometimes contains mucous.
- Other symptoms: Gassiness, abdominal cramps or pain.

Inflammatory Bowel Disease

Inflammatory bowel disease includes Crohn's disease and ulcerative colitis, both of which have chronic diarrhea as a symptom. Both are incurable chronic diseases of the digestive tract that may be treated with surgery or managed with medication.

- How long diarrhea lasts: More than four weeks.

- Triggered by: Not related to a specific food.
- Appearance: Blood or mucous in stool.
- Other symptoms: Abdominal pain, fever, weight loss, delayed growth in children.

Irritable Bowel Syndrome

Irritable bowel syndrome (IBS) describes chronic diarrhea, constipation, and abdominal pain that does not have a known disease as a cause. If you have been diagnosed with IBS, discuss the possibility of celiac disease with your doctor. The American College of Gastroenterology recommends that anyone diagnosed with IBS and diarrhea be tested for celiac disease.

- How long diarrhea lasts: At least six months.
- Triggered by: Not related to a specific food, though certain foods may aggravate symptoms.
- Appearance: Small, frequent stools.
- Other symptoms: Chronic abdominal bloating or distention, constipation. Pain is relieved by bowel movement.

Milk/Soy Protein Intolerance

Infants usually show signs of protein intolerance within a few months of birth. Some infants may react to food proteins present in breastmilk while others may react to formulas based on either cow's milk or soy.

- How long diarrhea lasts: More than two weeks.
- Triggered by: Dairy or soy products, sometimes by egg or other proteins.
- Symptoms appear within: two hours or more.
- Appearance: Streaks of mucous or blood.
- Other symptoms: Distended belly, crying, failure to thrive.

Medication

Some medications, in particular antibiotics and chemotherapy, can cause diarrhea, as well as laxatives containing magnesium. You may have a reaction to the medication itself or an additive, such as a flavoring. The medication can also alter the balance of bacteria in your gut, causing abdominal pain and diarrhea. Talk to your doctor if you have diarrhea after starting a new medication.

Genetics

There are congenital diarrheal disorders linked to specific genes. These disorders usually come on in the first few months of a child's life. They are often most common in certain populations, although congenital chloride diarrhea is one that appears worldwide.

Some other conditions that may display chronic diarrhea also tend to run in families, including celiac disease, some forms of lactose intolerance, and food allergies.

Lifestyle Risk Factors

A change to your diet, such as going on a mostly liquid diet, eating too much fruit or fiber, or eating spicy foods may lead to diarrhea.

Food poisoning, one of the leading causes of diarrhea, is often caused by improper food handling.

These four habits can reduce your risk of food poisoning, according to the Centers for Disease Control:

- Clean: Keep your kitchen clean, wash utensils and cutting boards with hot, soapy water.
- Separate: Raw meat, seafood, poultry, and eggs should be kept separate from other food. Use a separate cutting board for these.
- Cook: Use a food thermometer to ensure meat is cooked to an internal temperature that will kill the bacteria that cause food poisoning.
- Chill: Don't let food be in the danger zone between 40 °F and 140 °F for more than two hours. Thaw frozen foods in the refrigerator or microwave, not out on the counter. Be sure your refrigerator is kept below 40 °F.

The bacteria, viruses, and parasites that cause traveler's diarrhea and stomach flu are spread by contact with contaminated surfaces, food, and water. In the medical world, this is called the fecal-oral route. To reduce your risks, wash your hands well after using the bathroom, changing your child's diapers, and before eating. If you don't have access to soap and water, use an alcohol-based hand gel.

Never drink untreated water from a natural source, such as a stream. Even in developed countries, they can be contaminated with diarrhea-causing parasites spread by wildlife, such as Giardia. When traveling to areas that have a higher risk of contaminated water and food, drink only bottled water and don't use ice unless it is from bottled or purified water. Avoid uncooked vegetables and fruits (unless they can be peeled), raw shellfish, undercooked meat, and dairy products.

Sometimes the cause of diarrhea is taking too many laxatives or longer-term abuse of laxatives.

GASTROENTERITIS

Bacterial gastroenteritis is a type of gastroenteritis – a common condition sometimes called stomach flu or food poisoning.

Gastroenteritis is the result of irritation and inflammation in the stomach and intestines. It can have many different possible causes, including infection with a virus, such as rotavirus, bacteria or parasites. When caused by bacteria, it is called bacterial gastroenteritis.

The condition can affect adults and children. Symptoms of bacterial gastroenteritis can range from mild to severe and vary depending on the type of bacteria that have caused the infection. The main symptom is usually diarrhea. Other symptoms may include:

- Abdominal (belly) pain or cramps.
- Loss of appetite.
- Nausea and vomiting.
- Fever.
- In severe cases, blood in the stool (feces); if blood is present, medical attention should be sought as a matter of urgency.

Bacterial stomach flu can be caused by a number of different types of bacteria, which are most often spread through contaminated food and beverages. Depending on the source of infection, a large group of people may develop symptoms of bacterial gastroenteritis, or food poisoning, at the same time.

While it can be very unpleasant, most cases of bacterial gastroenteritis can be treated at home, and clear up within a few days without causing complications. Treatment for bacterial gastroenteritis typically involves:

- Resting at home.
- Drinking plenty of fluids to avoid dehydration.
- When necessary, taking medication to relieve nausea and diarrhea.

Antibiotics are generally only recommended for more severe types of bacterial gastroenteritis.

Symptoms of Bacterial Gastroenteritis

Signs and symptoms of bacterial gastroenteritis include:

- Diarrhea,

- Abdominal (belly) discomfort, pain or cramps,
- Bloating,
- Loss of appetite,
- Nausea and vomiting,
- Fever and chills,
- Body aches,
- Headache,
- Dizziness,
- Feeling of weakness,
- In severe cases, blood in the stool (feces).

Symptoms may be mild or more severe, and vary depending on the type of bacteria.

If symptoms are severe or do not improve within 2-3 days, it is important to contact a doctor without delay. Furthermore, medical attention should be sought immediately if:

- There are any indications of severe dehydration, such as passing little to no urine or dizziness that does not go away.
- There is blood, pus or a black color in the diarrhea.
- There is constant vomiting that makes it impossible to keep down fluids.
- There is very intense abdominal pain.
- There is a fever over 38 °C Celsius (101 °F), in adults or children.

In addition, a doctor should be contacted urgently if the affected person:

- Is pregnant,
- Is a very young child,
- Is an elderly person,
- Has a weakened immune system,
- Has a chronic condition such as inflammatory bowel disease or diabetes.

Causes of Bacterial Gastroenteritis

Stomach flu can be caused by many different bacteria, including:

- Salmonella,

- Campylobacter,
- Shigella,
- Escherichia coli (E.coli),
- Yersinia,
- Staphylococcus,
- Clostridium difficile.

In some cases, the bacteria may be transmitted directly from one person to another, but they are typically spread through contaminated food and water. Sources of bacterial stomach bugs include:

- Unhygienic food preparation, e.g. a cook not washing their hands after going to the toilet, or using the same cutting boards for both meat and salads.
- Raw or undercooked meat, eggs and fish.
- Unpasteurized dairy and juices.
- Inadequately-treated drinking water.

Bacterial food poisoning can also be contracted from food that is not stored properly, e.g. poor refrigeration that leads to food spoiling, as well as from raw fruits and vegetables that are not thoroughly washed in clean water. Contact with reptiles, birds and amphibians can also be a route of transmission for Salmonella bacteria.

Traveler's Diarrhea

Bacterial gastroenteritis is a common cause of traveler's diarrhea, a condition which frequently affects overseas travelers, particularly in developing countries. It is often the result of inadequate food hygiene at local restaurants and food outlets.

Bacterial Gastroenteritis caused by Exotoxins

Some cases of bacterial gastroenteritis are not caused by bacteria themselves, but instead by toxins that they release. The harmful byproducts of certain types of bacteria can contaminate food, causing a type of food poisoning when ingested. These byproducts are called exotoxins, and they usually cause symptoms within 12 hours of consuming the contaminated food. This type of gastroenteritis typically clears up within 36 hours.

Bacterial Gastroenteritis Treatment

Most cases of bacterial gastroenteritis clear up without specific medical treatment. As long as symptoms are not severe and there are no signs of dehydration, a person can generally treat bacterial gastroenteritis in an adult or child at home.

Home Remedies

The following home remedies and over-the-counter treatment approaches are recommended for mild cases of bacterial stomach flu:

- Bed rest,
- Consuming plenty of fluids in the form of water or oral rehydration drinks; at the very least, small sips should be taken between being sick or having bowel movements,
- Eating light meals when the appetite returns; plain foods like bread and rice may be helpful,
- Taking an antidiarrheal medication, e.g. loperamide, only when necessary and there is no fever, and no blood or mucus is present in the stool,
- Taking antiemetic (anti-nausea) medicine for nausea and vomiting, when necessary.

Good to know: In the past, the BRAT (Bananas, Rice, Applesauce, Toast) diet was often recommended for an upset stomach, e.g. one caused by gastroenteritis. However, because this bland diet lacks important nutrients like protein, it is now recommended that a regular, balanced diet be resumed as soon as possible for the person.

If a person is unable to keep any fluids down or shows signs of dehydration, they may need to be treated with an IV drip at a hospital.

Antibiotics for Bacterial Gastroenteritis

In mild cases of bacterial gastroenteritis, antibiotics are typically not considered necessary and will not be prescribed. Antibiotic medication is generally only recommended in severe or persistent cases of bacterial food poisoning, in some cases of traveler's diarrhea, in infections with C. difficile and in cases where the person has a weakened immune system or is otherwise at risk of complications. The type of antibiotics needed will depend on the specific type of bacteria, which may need to be determined with laboratory tests.

Complications of Bacterial Gastroenteritis

Generally, a bout of bacterial gastroenteritis clears up completely, without causing any complications. In a few cases, there may be complications. The risk of developing complications from bacterial gastroenteritis is highest in very young children, elderly people, and people who have a chronic condition such as diabetes, or who have a weakened immune system.

If stomach flu occurs in these groups of people, medical advice should be sought without delay.

Dehydration

Dehydration is the Most Common Complication of Bacterial Gastroenteritis. Dehydration may present with the following signs and symptoms in adults:

- Urinating less and dark colored urine,
- Dry mouth,
- Tiredness,
- Dizziness,
- Headache,
- Muscle cramps,
- Eyes appear sunken in the head,
- Weakness.

Additional signs which may help to identify dehydration in children include:

- Drowsiness or irritability,
- Increased heart rate,
- Fewer wet nappies,
- No tears when crying.

If signs and symptoms of dehydration do not go away with the replacement of lost fluids, or if a person is unable to keep fluids down, a doctor should be seen immediately. Dehydration can be extremely serious.

Other Complications

Other complications of bacterial stomach flu may include:

- Reduced effectiveness of medications, e.g. birth control pills or diabetes medicine.
- Temporary intolerance to lactose, the sugar in cow's milk.
- Irritable bowel syndrome.

Rare complications from bacterial gastroenteritis include:

- Reactive inflammation in other parts of the body, causing conditions like reactive arthritis.

- The infection spreading to other parts of the body, e.g. the bones, joints, gallbladder, or the tissue around the brain and spinal cord (meninges).
- Persistent diarrhea.
- Hemolytic uremic syndrome, a serious condition characterized by anemia and kidney failure.
- Bleeding in the gastrointestinal tract.
- Guillain-Barre syndrome, a serious condition in which the body's immune system mistakenly attacks the nerves.

Prevention of Bacterial Gastroenteritis

Good personal hygiene and food preparation hygiene, as well as safe food storage and the use of water only from clean, adequately-treated sources, can help to reduce the risk of developing bacterial gastroenteritis.

The following general precautions are recommended:

- Always wash your hands well with soap and water, or use an alcohol-based hand sanitizer, after using the toilet.
- Always wash your hands well before preparing food, and after handling any raw meat or fish.
- Wash fruit and vegetables in clean water before using them.
- Use different chopping boards for meat and vegetables, and clean them thoroughly with a disinfectant after use.
- Regularly clean kitchen work surfaces with a disinfectant.
- Cook all meat and fish well.
- Store food in clean containers at appropriate temperatures; throw out anything that seems to have spoiled.
- When traveling to other countries, drink only bottled water or water that has been boiled for at least 10 minutes; and only eat food that has been well cooked and fruit that can be peeled.

Bacterial gastroenteritis is contagious and can be spread easily. For this reason, it is important for people who have bacterial gastroenteritis to take precautions to avoid passing it on to other people. If you have bacterial gastroenteritis, in addition to the actions above, doing the following can help to prevent the spread of the infection:

- Clean the toilet and bathroom thoroughly with disinfectant on a daily basis.

- Avoid sharing cutlery, towels, clothing and linen; also wash your items separately in hot water, with bleach if possible.
- Avoid preparing food for other people for at least two days after the vomiting or diarrhea has cleared up.
- Stay off work or school, if possible, until at least two days after the vomiting or diarrhea has cleared up.
- Avoid swimming for two weeks after the vomiting or diarrhea has cleared up.

TOXOPLASMOSIS

Toxoplasma gondii was first described in 1908 by both Nicolle and Manceaux in the North African rodent, Ctenodactylus gondii and by Splendore in rabbits in Brazil. The parasite has since been recognized as a common infection in numerous warm-blooded animals, including humans. With approximately 2 billion people infected it is one of the most successful human parasites. The clinical significance of toxoplasmosis was first recognized in the 1920s in congenitally infected children presenting with hydrocephalus, retinochoroiditis and encephalitis. In the 1980s, T. gondii emerged as a major opportunistic infection in the setting of AIDS, presenting as a severe and potentially fatal encephalitis due to the reactivation of latent infections in the setting of HIV associated immune suppression. Toxoplasmosis is also a clinically important opportunistic infection in other immune suppressed individuals such as patients undergoing cancer treatment or organ transplant. Immune competent individuals infected with T. gondii are typically asymptomatic; however, infection has been associated with neuropsychriatric disorders suggesting that latent infection may have subtle neurological effects.

The Parasite and its Life Cycle

Toxoplasma gondii is an obligate intracellular protozoan parasite belonging to the Phylum Apicomplexa. There are three basic infectious life cycle stages of this parasite: sporozoites, tachyzoites and bradyzoites. Cats and other felidae serve as the definite host producing oocysts, an environmentally resistant form found in infected cat feces, transmitted by ingestion which causes infection transforming to tachyzoites. Bradyzoites are located primarily in tissue cysts in muscle and brain of the intermediate host and when ingested bradyzoites cause infection transforming to bradyzoites.Tachyzoites are the rapidly replicating form of this organism that disseminate throughout the body. Tachyzoites can transmit infection if they are injected into a host (e.g. laboratory accident or via blood transfusion). T. gondii sporozoites, tachyzoites and bradyzoites are all crescent-shaped, approximately 2 to 6μm in width and 4 to 8μm in length, with an anterior end containing secretory organelles, called the apical complex, that are used in invasion. Locomotion occurs via a process, called gliding motility, an actin-based based mechanism, which is essential for parasite migration through tissues, across biological barriers, in host cell invasion and egress. This form of 'gliding motility' is unique to the Apicomplexa and is an active target for drug development.

Cats are the definitive host in which schizogony (asexual reproduction) and gametogony (sexual reproduction) occur in the epithelial cells of the small intestine, leading to the production of unsporulated oocysts. These are shed in feces and maturation to infectious sporulated oocysts occurs in the environment. Infection in the cat can occur due to ingestion of bradyzoites, tachyzoites or oocysts; however, the prevalent period (i.e. the time to shedding of oocysts after infection) varies according to the life cycle stage ingested. The shortest prevalent period follows ingestion of tissue cysts (3-10 days) then tachyzoites (13 days) and the longest follows oocyst ingestion (≥ 18 days).

The tachyzoite is present during acute infection. Tachyzoites replicate every 6 to 8 hours in a process called endodyogeny.They replicate within an intracellular compartment termed the parasitophorous vacuole. This vacuole avoids fusion with the endolysosomal pathway of the host cell and is essential to intracellular growth of the parasite. The parasite induces profound affects in host gene expression, including molecules involved in signal transduction pathways and immune defenses, such as STAT pathways and apoptosis that facilitate the parasite survival in host cells.Tachyzoites are capable of infecting every type of nucleated cell and disseminate throughout the body. Due to the host immune response tachyzoites are eliminated from the host within a few weeks following acute infection.

The bradyzoite, a slower replicating form, differentiates from the tachyzoite stage and is prevalent in chronic infections.Bradyzoite replication does not lead to rupture of the host cell but rather, vacuoles containing bradyzoites mature into cysts that have a thick wall containing a parasite glycoprotein called CST1, which may be important in stability and protection of cysts from the host immune response. Cysts develop intracellularly reaching diameters of up to 100 μm, and can contain thousands of bradyzoites. Tissue cysts are located primarily in the grey matter and within neurons. Bradyzoites can be found in the brain as early as 3 days following infection. Cysts containing viable bradyzoites, persist for years and probably for the life of the host. Brain and muscle are the most common sites of the chronic, latent infection, although cysts have also been found in lung, liver, kidney and other visceral organs.

Transmission

Transmission occurs predominantly via ingestion of the tissue cysts, ingestion of oocysts from contaminated food or water, or congenital transmission. It is not yet clear which of these routes is the most important epidemiologically in transmission although this may vary depending upon culture and eating habits. Infections from tachyzoites in tissue transplants, blood products and unpasteurized milk may also occur.

Tissue Cyst Transmission

Tissue cysts are present in the muscle and brain of the intermediate hosts. Cats acquire toxoplasmosis by eating infected prey (rodents, birds) wherein the enteroepithelial cycle, sexual reproduction and production of oocysts occurs. Wild carnivorous animals, including bears, fox, raccon and skunks, also can acquire toxoplasmosis via ingestion of tissue cysts. Man can acquire the infection via consumption of undercooked contaminated meat such as pork or lamb. When a tissue cyst is ingested the cyst wall is disrupted by proteolytic enzymes in the stomach and the released bradyzoites are resistant to proteolytic digestion allowing them to survive digestion initiating infection in the small intestine.

Oocyst Transmission

Infected cats excrete oocysts in large numbers (up to 10 million cysts during a single day) with excretion occurring for up to 2 weeks following infection. Once shed, the oocyst sporulates in 1 to 5 days, becoming infective and may remain infective for more than 1 year in unfrozen, moist soil. Humans can become infected from food or water contaminated with oocysts. Eating of unwashed raw vegetables or fruits or ingestion of water contaminated with oocysts, have been identified as important risk factors. Circumstantial evidence suggests that oocyst-induced infections in humans are clinically more severe than tissue cyst-acquired infections. While water-borne infections were once thought to be rare, the widespread infection of marine mammals in the USA as well as outbreaks linked to contamination of a municipal water reservoir in Canada by wild felids has led to the recognition that toxoplasmosis can be a water-borne disease. Because of the short duration of oocyst shedding, the passing of non-infective oocysts, and the fastidious nature of cats, direct contact with cats is not thought to be a primary risk for human infection. Oocysts have been found to survive in seawater for up to 6 months, suggesting that in coastal marine environments they could be a source of infection via transport hosts. In support of this idea, it has recently been demonstrated that oocysts can survive in filter feeders such as anchovies and sardines, and remain infectious inside their alimentary canal.

Congenital Transmission

Congenital transmission occurs during acute toxoplasmosis in a seronegative mother when tachyzoites present in the blood may cross the placenta and infect the fetus. The stage of pregnancy in which maternal toxoplasmosis is acquired is an important factor in the frequency of transmission and severity of congenital infection. During the first trimester, transmission is relatively low (< 20%) but increases to near 80% by the end of the pregnancy. Early in gestation cases are severe with infection resulting in spontaneous abortion, hydrocephaly and mental retardation while during the last trimester, transmission rates are highest but the majority of these cases are subclinical resulting in asymptomatic infections or recurrent chorioretinitis throughout early adulthood that can lead to vision problems and potentially blindness. In most populations the overall incidence of congenital toxoplasmosis is from 1 in 1000 to 1 in 10,000 live births. A recent study measuring vertical transmission rates in humans using a PCR detection of the umbilical cord tissue, obtained at birth, found transmission rates of 19.8%. In studies in sheep this rate was 65% and in experimental mouse models 75%. Overall, this suggests that the human vertical transmission rates may be underreported when they are based upon serological approaches.

Immune Response

Resistance to T. gondii is mediated primarily by a Th-1 type cell-mediated immune response that depends upon the production of interleukin-12 (IL-12) and interferon-γ (IFN-γ). Synthesis of IL-12 is stimulated by dendritic cells (DCs), macrophages and neutrophils, which are infected by the parasite, early in infection and induces IFN-γ synthesis by NK and T lymphocytes. IFN-γ is the main cytokine controlling both the acute and chronic stages of infection. IFN-γ activates anti-parasitic effector mechanisms in both hemopoietic cells, such as macrophages, and non-hemopoietic cells such as endothelial cells, fibroblasts, and astrocytes; both cell populations have been found to be necessary for control of the parasite. In immunodeficient hosts, such as AIDS patients, when $CD4^+$ T cells fall and correspondingly, IFN-γ levels decline, reactivation of the chronic infection in the brain results. In most healthy (immunocompetant) hosts, the immune response controls replication during the acute infection and due to immune response and/or physiological stress, the tachyzoites differentiate into bradyzoites forming cysts, primarily in the brain and muscle, that then persist throughout the host's life.

The immune response to T. gondii requires both innate and adaptive immune responses, involving a coordinated series of cellular interactions between the parasite, enterocytes, monocytes, DCs, macrophages, NK cells, and neutrophils. DCs play a central role in the stimulation of the innate immune response and the initiation of the adaptive immune response. IFN-γ dependent cell immunity plays a major role in resistance against both acute and chronic T. gondii infection, with IFN-γ activated effector cells in both hemopoietic and non-hemopoietic cell compartments controlling T. gondii via several antimicrobial mechanisms. CD8+ T cells are crucial for protection against T. gondii, The development of imaging techniques such as bioluminescent imaging, confocal and multiphoton microscopy, in combination with fluorescent and bioluminescence markers have permitted visualization of parasites and immune cells in real-time and are providing new insights into the molecular nature of the host-parasite interactions.

Innate Immune Responses

Intermediate hosts become infected with T. gondii by ingestion of oocysts or tissue cysts, where the parasite invades the enterocytes in the small intestine and replicates. Parasite replication in the intestine leads to host cell lysis and release of tachyzoites.

Parasites can also traverse epithelium, powered by gliding motility, without the disruption of the endothelial layer and infect the lamina propria directly. Enterocytes infected with parasites secrete chemokines that recruit dendritic cells (DCs) in the lamina propria. Within the lamina propria, the parasite can either invade DCs or macrophages or enter directly into the lymphatic or circulatory systems. Infected DC and macrophages can disseminate the infection to distant sites, including the brain. Macrophages and DC produce IL-12 after parasite infection, which then triggers IFN-γ synthesis by NK, NKT, and T cells. Intraepithelial cells within the lamina propria also contribute to innate immunity in the lamina propria.

Role of Monocytes and Neutrophils

Monocytes play a key role in mucosal immunity to T. gondii. Inflammatory monocytes are recruited from the bone marrow to the site of infection in the gut. Inflammatory monocytes can restrict the growth of T. gondii through the production of nitric oxide (NO) and recent studies indicate monocytes are necessary in control of T. gondii proliferation in the lamina propria via a NO-dependent mechanism. Neutrophils are also one of the first cell types to arrive at the site of infection with T. gondii. They secrete IL-12 and can directly kill the parasite via oxygen-dependent and independent mechanisms and thus have also been touted as having a protective role in mucosal immunity. More recent studies however have shown neutrophils to be involved in adverse pathological events.

Role of Dendritic Cells (DCs)

Dendritic cells (DCs) play a central role in orchestrating the innate immune response and bridging innate and adaptive immunity. DCs are the initial antigen-presenting cell at the site of infection and they are either directly infected by the parasite where they multiply and are processed for antigen presentation, or ingest apoptotic enterocytes that are then digested and processed for antigen presentation. While macrophages and neutrophils are also infected with T. gondii and can secrete IL-12, recent evidence indicates DC production of IL-12, but not macrophages or neutrophils, is essential for IFN-γ activation of NK cells and recruitment of inflammatory monocytes. Both conventional and plamacytoid DC subsets are activated by T. gondii infection, with the plasmacytoid DC subset inducing high levels of IL-12, upregulating MHC I and MHC II, and accessory molecules, such as CD86, indicating their ability to prime CD4+ T and suggesting the plasmacytoid DC subpopulation may be particularly important in the initial stages of the immune response to T. gondii.

DCs are also involved in priming CD8+ T cells and shaping the early T cell response. Cognate formation between DCs and CD8+ T cells occurs in the draining lymph nodes early in the immune response. The DC/CD8+ interaction only occurs only in the presence of cognate antigen and not infected DCs, suggesting that cross presentation by bystander DCs rather than infected DCs is an important route of antigen presentation during acute toxoplasmosis. An enhanced migratory response is induced in infected DCs, which likely facilitates parasite dissemination and DCs trafficking to the draining lymph nodes. Additionally, the infected DC also contribute to parasite dissemination in a strain-specific manner with Type II avirulent strains more efficiently exploiting DCs for parasite dissemination than the virulent, Type I strains. Avirulent strains induce a more robust activation of DCs at the site of infection and the draining lymph node than the virulent strains, further indicating the role of DCs in shaping the immune response to T. gondii.

Role of TLR's in Innate Immune Response

Several Toll like receptors (TLRs) function in recognition of T. gondii. TLR11 is required for DC production of IL-12 production and thus is important in induction of innate immunity. TLR11 recognizes the parasite protein profilin, which is an actin-binding protein that is used in gliding motility, involved in tissue migration, host cell invasion and egress. In addition to triggering host immune response, TLR11 may also be involved in the prevention of immunopathology via regulating IFN-γ secretion by NK cells. TLR2 also contributes to the host response to T. gondii, inducing macrophage activation and NO production and secretion of the chemokine, CCL2. The ligands for TLR2 are glycophosphatidylinositol (GPI) anchors of the parasite. TLR4 also recognizes GPIs of the parasite, with some suggestion that TL2 and TLR4 may work cooperatively during T. gondii infection. These studies indicate TLRs may be involved in a cooperative regulation of the innate immune response against T. gondii.

Adaptive Immune Response

The adaptive immune response is initiated by antigen-presenting cells (APCs), which recognize that the host is infected through pattern recognition receptors such as TLRs, stimulating secretion of IL-12, which induces IFN-γ production by NK cells and $CD4^+$ and $CD8^+$ T cells. Activated $CD4^+$ cells produce IL-2, a T-cell mitogen, which in conjunction with IFN-γ, results in large numbers of parasite-specific$CD4^+$and CD8+ T cells that produce IFN-γ at the sites of parasite invasion. IFN-γ signaling, in a STAT-1 dependent pathway, leads to the generation of anti-parasitic effector mechanisms such as nitric oxide (NO), reactive oxygen intermediates (ROI) and the immunity-regulated GTPases (IRGs). CD8+ T cells play a crucial role in the protective immunity to T. gondii both via the production of IFN-γ, which activates anti-microbial killing mechanisms but also via direct killing of the parasite via a perforin-dependent pathway. The chronic infection in the brain is controlled by CD8+ T cells and the resident glial cells, microglia and astrocytes, which can act as immune effector cells in the brain.

Role of CD8+ T Cells

CD8+ T cells are a key mediator of the adaptive immune response to T. gondii. The role of CD8+ T cells in the protective response to T. gondii was established in early studies in which transfer of CD8+ T cells from immunized mice provided protection to naïve mice challenged with T. gondii and correspondingly protection was ablated by the depletion of CD8+ T cells. CD8+ T cells are the principle T lymphocyte population involved in the control of the chronic infection and prevention of reactivated infection in the brain.CD4+ T cells, are not required for induction or maintenance of CD8+ T cells but are necessary to regulate the functional activity of intracerebral CD8+ T cells.

CD8+ T cells are activated via presentation of major histocompatibility complex class I (MHC-I)-restricted parasite antigens. Activation of T cells occurs via direct interaction with infected host cells as opposed to cross-presentation as has been found in the initial priming of CD8+ T cells by DC in the lymph nodes. Professional APCs, such as dendritic cells and macrophages, are involved as in direct antigen presentation, but non-professional APCs, such as endothelial cells, epithelial cells and astrocytes, are equally as efficient in activating CD8+ T cells in this MHC-I restricted response.

The anti-parasitic effector mechanisms mediated by activated CD8+ T cells include production of cytokines, the predominant of which is IFN-γ, which stimulates anti-parasitic IFN-γ effector mechanisms in a variety of cell types including macrophages, microglia, and astrocytes. The importance of IFN-γ-mediated mechanism is illustrated by the loss of protection against both the acute and chronic stages of infection when IFN-γ is deficient or absent. CD8+ T cells are also known to produce the anti-inflammatory cytokines, IL-17 and IL-27, both of which are associated with down regulation of the inflammatory response to T. gondii during infection.

Activated CD8+ T cells can also mediate cytotoxic activity (CTL) against Toxoplasma-infected cells in the context of MHC presenting parasite antigen on their cell surface. The ability of parasite-specific CD8+ T cells to lyse Toxoplasma-infected cells has been demonstrated in in vitro studies. In vivo studies using mice deficient in the cytotoxic molecule, perforin, demonstrated the CTL response did not affect resistance to the acute stage but was necessary for resistance to the chronic infection. CD8+ T cells may also be able to eliminate Toxoplasma cysts in the brain. Surprisingly the CD8+ T cells did not require production of IFN-γ but did require perforin-mediated cytotoxic activity.

IFN-γ Effector Mechanisms and IFN-γ Effector Cells

IFN-γ activated macrophages acquire anti-microbicidal activity that is largely dependent upon the production of reactive nitrogen intermediates. The Interferon Response Genes (IRGs), are a family of IFN-γ induced proteins, that have recently been recognized as an important resistance system to target vacuolar pathogens including T.gondii. In macrophages the IRG mechanism induces disruption of the Toxoplasma vacuole leading to degradation of intracellular parasites via autophagomal delivery to the lysosomes. The IRG mechanism involves a coordinated loading of IRG GTPases on the Toxoplasma vacuole. Virulent strains of T. gondii evade vacuolar disruption by IRG proteins via phosphorylation of several IRG proteins, via the parasite kinase, ROP18, thereby inhibiting the accumulation of IRG proteins on the vacuole. IFN-γ induced damage to the T. gondii vacuole can also occur via an autophagosome-independent process involving the autophagy protein, Atg5.

IFN-γ also induces anti-Toxoplasma activity in non-phagocytic cells such as fibroblasts, epithelial cells, endothelial cells and astrocytes and in these nonphagocytic cells, the mechanism of IFNγ inhibition is via nitric oxide independent mechanisms. The known mechanisms of IFNγ inhibition in non-phagocytic cells includes induction of indoleamine 2,3-dioxygenase (IDO), which degrades intracellular tryptophan, increases in reactive oxygen species, and iron deprivation. The IRG mediated parasite degradation has also been found in fibroblasts and astrocytes.

Intracerebral Immune Response

Infected DCs and macrophages can carry the parasite into the brain. The parasite can infect microglial, astrocytes and neurons during acute Toxoplasmic encephalitis (TE) and then persist in primarily in neurons during chronic TE. CD8+ and CD4+ T cells are recruited in the brain in acute TE. The recruitment of inflammatory cells into the brain is regulated by cytokines and chemokines. Trafficking of CD8+ T cells into the brain depends upon IFN-γ induction of VCAM1 on endothelial cells that binds to an integrin on CD8 T cells, guiding CD8 T cells into the brain. Both IFN-γ activated astrocytes and microglia produce chemokines in response to parasite infection that guide the inflammatory cells to the site of infection.

CD4+ and CD8+ T cells increase in the brain during the first weeks following infection and during the transition from acute to chronic TE gradually decline with CD8+ T cells persisting at low levels in the brain during chronic TE. More recent studies using Ag-specific CD8+ T cells, have found that recruited CD8 T cells persist in the brains of infected mice in the presence of ongoing antigen recognition. In this study cerebral CD8+ T cells, made transient contact with granuloma-like structures containing parasites and with individual macrophages (CD11b+ cells) but surprisingly not with intact Ag-bearing cysts, infected neurons containing parasites or the cysts, or individual neurons or astrocytes, indicating macrophages serve as APCs in the brain, while astrocytes and neurons may not. Live cell imaging, using 2-photon microscopy, observing CD8+ T cells in the brain revealed 2 distinct populations of cells, one with a constrained pattern of migration and one with a highly migratory subset, the proportion of which changed over time and varied with antigen availability. Unexpectedly, CD8 T cells moved along infection induced reticular fibers, to guide CD8 T cells to the site of infection. These studies have revealed features of the intracerebral immune response and behavior of CD8+ T cells that were unanticipated and not fully understood, but that likely will be the focus of future studies.

Role of Microglial Cells

Microglia are the resident macrophage population of the central nervous system and are rapidly activated in murine Toxoplasmic Encephalitis (TE). Activated microglia express elevated levels of the cytokines, IL-12, IL-1β1, TNFα, and the cell surface ligands, MHC I and II and ICAM-1, indicating their role in recruitment and activation of T cells in the brain and ability to act as an APC in the brain. CD8+ T cells regulate cytokine production of microglia, while cell surface molecules are less dependent upon T cells. Macrophages and microglia can also be source of IFN-γ in the brain and produce IFN-γ independently of T cells during acute toxoplasmosis in the brain. Microglial production of IFN-γ independent of T cells, may be important in the early stages of tachyzoite proliferation in the brain, serving to limiting parasite growth, prior to the recruitment of CD8 T cells into the brain. Microglia can also serve as IFN-γ activated immune effector cells in the brain. The mechanism of IFNγ inhibition of T. gondii infection in murine microglia is primarily mediated via inducible nitric oxide synthase (iNOS) and the production of nitric oxide. Nitric oxide (NO) is believed to be toxoplasmacidal, and in sufficient amounts may produce stasis and/or intracellular killing of parasites. Activated microglial cells can inhibit growth of the parasite via CD40-autophagy pathway in microglia. Microglia can also function in the down-regulation of the intracerebral immune response via the secretion of TGF-β which prevents neuron degeneration, indicating microglial play an important neuroprotective function in the brain in chronic TE.

Role of Astrocytes

Astrocytes are also important immune effector cells in the brain, controlling parasite growth and immunopathological responses to the chronic infection. IFN-γ activated astrocytes significantly inhibit the growth of T. gondii. IFN-γ inhibition in astrocytes in mice is via the IRG-mediated mechanism while inhibition in human astrocytes is via nitric oxide. Astrocytes, are efficient in MHC I presentation, indicating they may be an important antigen presenting cell controlling T. gondii in the brain, capable of stimulating CD8+ T cells. Soluble factors produced by secreted by infected astrocytes down modulate NO-production by IFN-γ activated microglia and prevent neuronal degeneration. Finally, astrocytes have been shown to be essential for control of Toxoplasma encephalitis, serving to contain sites of parasite replication and the inflammatory lesions in the brain.

Humoral Immune Response

Although the predominant protective response to T. gondii is a Type I-mediated cell mediated immune response, humoral immunity also participates in controlling T. gondii. Antibodies are produced in response to T. gondii infection and individuals with a chronic infection maintain the Toxoplasma-antibodies for life. Evidence from antibody deficient mice (μ−/μ−)indicate while antibody did not affect the acute infection, antibodies did play a role in limiting tachyzoite replication in mice, especially in the lung and in the brain. Additionally it is known that B cells can modulate CD4 and CD8 responses and likewise, B cells have been found acquire the ability to amplify IFN-γ production by CD4+ and CD8+ T cells during the course of the Th1 inflammatory response to T. gondii infection.

Immunoregulation during T. Gondii Infection

IL-10 is a cytokine with anti-inflammatory properties including counteracting the function of Th1 lymphocytes. IL-10 plays a vital role in controlling the inflammatory response during chronic phase of T. gondii infection. IL-10 inhibits the production of IL-12, IFN-γ, TNF-γ and IL-6 from macrophages and microglia in the brain. Macrophages and CD4+ T cells are the major sources of IL-10 in the brain. Depletion of IL-10 in the brain results in an increase in CD4+ T cells and macrophages and a lethal inflammatory response, indicating the critical role of IL-10 to limit inflammation in the brain during TE. The down regulation of the Type 1-immune response in the brain may facilitate the persistence of the parasite in the brain resulting in the chronic infection. Host protective IL-10 derives in an autocrine manner from IFN-γ producing Tbet(+) Foxpβ(neg) Th1 cells, indicating that IL-10 may be an important self-regulatory function of CD4+ T lymphocytes.

SCOMBROID FISH POISONING

Scombroid fish poisoning, otherwise known as histamine fish poisoning (HFP), is the most common seafood poisoning in the United States related to the improper storage of fish. Because of misdiagnosis and other factors, HFP is undoubtedly underreported. Enteric bacteria (Escherichia coli, Proteus morganii, Morganella morganii, and Proteus vulgaris) act on the flesh of the poorly maintained fish to produce elevated histamine levels (as well as other bioactive amines) through the breakdown of the amino acid histadine. These organisms are part of the normal flora of certain fish and not thought to be the contaminants. Histamine production is correlated to the histadine content of fish, bacterial histadine decarboxylase, and environmental factors. Histadine levels vary from 1 g/kg in herring to as much as 15 g/kg in tuna.Unspoiled, fresh fish do contain small amounts of histamine (<0.1 mg/100 g), but do not cause HFP.

Reservoirs

The vast majority of fish causing HFP are members of the families Scombroidae and Scomberesocidae (hence the name scombroid fish poisoning), which include tuna, mackerel, skipjack, bonita, suary, and seerfish. Several nonscombroid fish are also linked to HFP, including bluefish, mahi-mahi, herring, sardines, and anchovies. These fish are nontoxic when afresh, can still appear very normal with the development of toxicity, and have no putrid odor. Other species that have also been affected are of little concern because they are not commonly eaten. The fish that are affected are thought to be related closely to specific endogenous bacteria. Interestingly, dark meat, a major component of the muscle of these fish, seems to be more affected. Improper storage of the fish at insufficient low temperatures appears to initiate the bacterial growth. Freezing, cooking, smoking, curing, or canning does not eliminate the potential toxins but destroys the bacteria that produce the toxins.

Foodborne Outbreaks

Outbreaks are worldwide and are found in places where potentially spoiled, improperly handled scombroid-like fish species are eaten. Of note is that HFP occurs with the ingestion of canned tuna, a mainstay of the American diet. Hence, wherever a can of tuna are sent creates the potential for HFP. Recreational fishermen tend to have less stringent habits and regulations for storing their catch and are hence more likely to end up with fish capable of producing HFP. The Board of Health in New York City reports approx. 70 cases of HFP per year, seen in emergency rooms or clinics, but clearly there are many more cases which are attested to a simple virus or food allergy (Dr. Herbert Schaumberg, personal communications). Last spring, there were 20 unreported HFP cases at Albert Einstein Medical Center alone after the Pediatric Department's farewell to the residents' banquet (Dr. Herbert Schaumberg, personal communications).

Pathogenicity (Virulence Factors)

HFP is generally associated with high histamine levels (>50 mg/100 g) in spoiled fish that have high histadine content. Histamine is endogenous to mast cells and basophils usually only having an effect when released in large amounts in response to an allergic reaction. Histamine has cell membrane receptors principally in skin, hematological, gastrointestinal, and respiratory systems. Interestingly, consuming spoiled fish with histamine is more poisonous than taking an equivalent amount of pure histamine orally), suggesting that there are other "scombroid toxins" acting with histamine. The histamine-potentiator hypothesis is based on the observation that absorption, metabolism, and/or potency of one biogenic amine might be altered in the presence of another amine. The biogenic amines putrescine and cadaverine are found in large quantities in toxic fish and in significantly lower levels in nontoxic fish. These amines, when given in higher ratios relative to histamine in toxic fish, have been shown to potentiate the effect of histamine in laboratory animals. Some amines, such as cadaverine and putrescine along with a number of other tested chemicals found in spoiled tuna, may competitively inhibit diamine oxidase (DAO), the main histamine catabolizing enzyme in the intestinal tract, and histamine methyl transferase (HMT). The toxicity of histamine can be potentiated by cadaverine and putrescinein guinea pigs. A number of drugs inhibit DAO including iproniazid, isoniazid,pargyline, aminoguanidine, phenelzine, and tranylcypromine, which could predispose one to HFP. Serotonin, tryptamine, and phenformin are good competitive inhibitors, whereas cimetidine and pheniprazine are noncompetitive inhibitors. Other antihistaminic drugs such as promethazine are less powerful inhibitors. Certain foods can inhibit DAO (alcohol and tyramine-containing foods such as strong cheeses) likely predisposing one to HFP as well. HMT is inhibited by analogs of methylmethionine such as adenosyl-homocysteine, antimalarial drugs, and numerous agonists and antagonists of histamine receptors. It should be noted that histamine has been implicated in the pathogenesis of migraine, the intolerance of which may be based on a deficiency in the DAO enzyme. A diet excluding alcohol and tyramine-containing foods is helpful for many patients with histamine-induced headaches. Regarding this fact, a worse reaction to HFP is expected if someone ingests wine with scombroid-laden fish.

The "barrier disruption hypothesis," first proposed by Parrot and Nicot, states that potentiators might interfere with mucin, which is known to bind histamine, thought to be an essential event in preventing the intestinal absorption of histamine and hence increases histamine absorption. It has also been postulated that histamine may be released endogenously by an unknown scombroid toxin. Urocanic acid, an imidazole compound and a histadine metabolite of spoiling fish, has been shown to induce histamine in vivo in mice and to degranulate mast cells in human skin organ cultures. The role of endogenous histamine release in HFP has not been established.

The sensitive polymerase chain reaction technique for detecting M. morganii, which was found to be present in fresh fish including mackerel, sardine, and albacore. The gill and skin were the main harbor sites of the bacterium in this study. Conventional culturing techniques did not detect the bacterium. This study demonstrates that histamine-forming bacteria are endogenous to these fish and are likely to proliferate under the suitable conditions. These findings also emphasize the importance of proper handling of fish, including refrigeration, to prevent HFP.

Clinical Characteristics

Fish affected by scombroid tastes "peppery," which is thought to be related to elevated histamine levels. Headache, a burning sensation in mouth and oropharynx, and nausea are common neurological symptoms of HFP. Other symptoms include diffuse erythema (especially flushing of the head, neck, and upper torso), headache, vomiting, diarrhea, and abdominal cramping. Occasional arrhythmias, hypotension, bronchospasm, and cardiovascular collapse also occur. Anxiety has also been reported as a prominent symptom with HFP. Generally, HFP is not a life-threatening condition, but in those patients with serious preexisting conditions, notable respiratory and cardiac complications do occur. The symptoms of HFP usually last for 8–12 h and can come on within minutes of ingestion of a toxic meal.

With HFP, there is a remarkably wide spectrum of clinical symptoms (mild to severe). Symptoms can be confused with Salmonella infection as well as food allergy. The usual HFP-associated fish species cause HFP much more frequently than they cause true allergic reaction. The clinical setting of the poisoning shows evidence to the etiology as if only one person is affected by a meal when several individuals have eaten the same fish, sheds doubt on the diagnosis of HFP. Nevertheless, the concentration of the histamine may not be evenly distributed through the fish and, hence, not everyone who eats the fish will necessarily get sick. In fact, food allergies to most of the common fish involved with HFP are rare.

Although the diagnosis can be confirmed by detecting histamine in contaminated fish, remnants of the meal, or similar fish from the same source as well as other biogenic amines (16) and urocanic acid, the diagnosis is normally made by the history and clinical presentation. There are numerous methods for the detection of histadine-decarboxylating bacteria that can be performed with specific media that are selective for this enzyme although some of these media have given false positive results. The biochemical methods for detecting histamine and other biogenic amines are extensive as well as controversial and are well outlined by Lehane and Olley. Measurement of plasma histamine levels is not routinely available in most community-based hospitals. Measuring human serum levels requires that histamine is measured within 4 h of the ingestion of the fish.

Other diagnoses to be considered with HFP are severe migraine, intracranial hemorrhage, and pheochromocytoma. HFP may resemble an immunoglobulin E (IgE)-mediated allergic reaction; however, HFP is a foodborne intoxication related to elevated histamine levels, and therefore the patient can safely eat the same type of fish again with impunity.

FOODBORNE CAMPYLOBACTER

Bacteria belonging to the genus Campylobacter are Gram-negative, spiral, non-spore forming rods that form spherical or coccoid bodies in older cultures. They are between 0.2 to 0.9 microns wide and 0.5 to 5 microns long, are motile and usually move with a polar unsheathed flagellum at one or both ends, and are microaerobic with a respiratory-type metabolism, although there are some that grow aerobically or anaerobically. The sequence of the genome of Campylobacter jejuni NCTC11168 was originally published in 2000.There is considerable variation between strains and re-annotation of the C. jejuni genome published in 2006 revealed that the complete sequence is 1,641,481 bp in length with 25 polymorphic regions. Moreover, new information for 1,450 of thc original 1,654 coding sequences revealed changes corresponding to over 300 product functions. The infectious diseases caused by members of the bacterial genus Campylobacter are called campylobacteriosis. Currently Campylobacter jejuni and Campylobacter coli are considered to be the most important enteropathogens among Campylobacter spp. The rate of Campylobacter infections are increasing worldwide, exceeding shigellosis.

Campylobacter Historical Perspective

From a historical standpoint, the first report of Campylobacter is believed to have been made in 1886 by Theodore Escherich, who observed and described a non-cultural spiral shaped bacteria, which he found in the colon of children with an enteric disease called "cholera infantum". Campylobacter was identified February 2, 1906 by two British veterinarians, John McFadyean and Stewart Stockman who reported the presence of "large numbers of a peculiar organisms" in Loeffler's blue-stained smears of uterine mucus from a pregnant ewe. In 1927, a group of vibrio-like bacteria was found in the feces of cattle with diarrhea. They were described by Theobold Smith and Marion Orcutt. In 1931, Jones and coworkers showed a relationship between the microaerophillic vibrios and bovine dysentery, and the organism was eventually called Vibrio jejuni. The first well documented incident of Campylobacter infection took place in Illinois in 1938.The case involved a milk-borne outbreak of diarrhea that affected 355 inmates in two adjacent state institutions. In 1944, Doyle isolated another vibrio from the feces of pigs with diarrhea and classified it as Vibrio coli. Because of their low DNA base (low guanine and cytosine) composition, non-fermentative metabolism and their microaerophillic growth requirements the genus Campylobacter was proposed by Seabald and Vernon in 1963, distinguishing them from the Vibrio spp. There are two subspecies recognized within C. jejuni, C. jejuni subspecies jejuni and C. jejuni subspecies doylei. Strains of C. doylei differ from C. jejuni biochemically in its inability to reduce nitrate and variable growth at 42 °C. Campylobacter coli and C. jejuni differ biochemically in their ability to hydrolyze hippurate. Campylobacter coli cannot hydrolyze hippurate and there are some C. jejuni subspecies that are hippurate negative.

Human Infections

Campylobacter are a leading bacterial cause of zoonotic disease worldwide, affecting approximately 1% of the human population in Europe each year and infecting 13 of every 100,000 persons in the United States annually. Clinical manifestations of campylobacteriosis are largely undistinguishable from other bacterial gut infections and include fever, abdominal cramping, and diarrhea, with or without blood in stools. These symptoms can last from three days to one week. In severe cases of C. jejuni infection, individuals may exhibit fever, abdominal cramps and diarrhea that contain blood and leukocytes, or they may develop post infection complications associated with Guillain Barré Syndrome or Miller Fischer Syndrome. Miller Fischer Syndrome is a subform of Guillain Barré Syndome characterized by areflexia, ataxia and opthalmoplegia. Campylobacter can also cause post-infection complications associated with acquired immune-mediated neuropathies such as Guillian Barré Syndrome or Miller Fischer Syndrome]. In the developed world, most human Campylobacter infections are caused by Campylobacter jejuni; however, 4% of clinically confirmed cases in one study were attributed to Campylobacter coli. Campylobacteriosis can occur in all age groups and the cost of human illness in the United States is estimated at $1.3 to 6.8 billion dollars annually.

In March of 2013 the Centers for Disease Control and Prevention (CDC) reported that in 2012 there was a 14% increase in the incidence of food-borne illness caused by Campylobacter jejuni. In 2005, the CDC estimated that the number of campylobacteriosis cases in the United States was about 1 million per year making it the second leading cause of laboratory confirmed cases of foodborne illness in the United States thus resulting in an estimated 12.72 cases per 100,000 people. In 2009, the Foodborne Disease Active Surveillance Network (FoodNet) of the CDC estimated that the number of infections by Campylobacter was a total of 6,033, or 13.02 per 100,000 people. In the last five years the European Food Safety Authority (EFSA) and the European Centre for Disease Prevention and Control (ECDC) reported that campylobacteriosis has become the most often reported zoonosis in the European Union followed by salmonellosis and yersinosis. More than 200,000 confirmed cases of campylobacteriosis were reported in 24 of the member states of the European Union at a rate of 45.2 cases per 100,000 people. In 2010 New Zealand reported the highest national campylobacteriosis rate, which peaked in May 2006 at 400 per 100,000 population. In general, developing countries do not have national surveillance programs for campylobacteriosis; consequently, there are no case incidence values in terms of population density. Most of the estimates of incidence are done by laboratories where the surveillance is based on pathogens responsible for diarrhea.

Campylobacter isolation rates in developing countries range from 5 to 20%. Most of the data was collected by the World Health Organization (WHO), which along with the Canadian Public Health Service have provided financial support to developing countries for epidemiologic studies. There is a large disparity in the incidence of campylobacteriosis in developing countries versus that of developed countries. High numbers of children in developing countries are affected by Campylobacter infections. Community based studies in developing countries have estimated that 60,000 per 100,000 children <5 years of age. This data suggests that campylobacteriosis is a pediatric disease in developing countries. In developed countries, more than 90% of human campylobacteriosis cases occur during the summer because of undercooked meats from outdoor cooking facilities. People of all ages are affected, but particularly children less than 4 years of age and young adults 15–44 years of age.

The infectious dose of Campylobacter jejuni for humans is estimated to be low; between 500–800 organisms. The dosage was estimated in 1981 as a result of a human experiment in which a British medical doctor, Robinson, swallowed 500 organisms of known serotype in 180 mL of pasteurized milk. The findings from this self-inflicted experiment, which reportedly satisfied the criteria of Koch's Postulates and the findings of additional human experiments, verified a dosage and a mechanism of C. jejuni human infection.

Campylobacter are also recognized as reservoirs for antimicrobial resistance genes that potentially can be exchanged between other pathogenic and commensal bacteria. Antibiotics have been used for decades in food production animals to control, prevent, and treat infections and to enhance growth. This usage has caused an increased resistance to multiple antibiotics by members of Campylobacter spp. in food production animals and environments. The development of antimicrobial resistance in Campylobacter spp. is a serious threat to human health. Worldwide, there has been a rapid increase in the proportion of Campylobacter strains resistant to antimicrobial agents. Campylobacter jejuni and C. coli are almost intrinsically resistant to penicillians, cephalosporins (with exception of a few 3rd generation cephalosporins), trimethoprim, sulfamethoxazole, rifampicin and vancomycin. For patients infected with Campylobacter spp., the prognosis is that most will recover without specific treatments other than replacement of fluid and electrolytes, however, in severe cases antibiotics such as macrolides and fluroquinolones are generally administered. It is believed that increasing resistance to fluroquinolones and erythromycin by C. jejuni and C. coli might compromise the effectiveness of these treatments.

Campylobacter Metabolism

Although there has been a significant amount of knowledge elucidated about the Campylobacter genus, there is still more to learn about Campylobacter's metabolic processes. For instance, the process of nutrient acquisition by Campylobacter has not been completely elucidated but it is apparent this process is flexible enough to allow survival in the environment (water, food or feces) before host ingestion. Unlike many other gut bacteria, however, Campylobacter are limited in their ability to conserve energy for growth and maintenance via fermentation of carbohydrates. These bacteria lack 6-phophofructokinase, which is a key enzyme in energy metabolism, although a novel L-fucose pathway has been elucidated by Muraoka and Zhang and Stahl. Moreover, Campylobacter are further limited in their ability to conserve energy for growth via fermentation of carbohydrates due to the absence an active phosphoenolpyruvate dependent phosphotransferase system, which would function to transport and phosphorylate sugars simultaneously. Campylobacter can conserve energy via respiration, oxidizing hydrogen and formate for the reduction of the electron acceptors (fumarate, nitrate, sulfites) and, if at low concentrations, oxygen, to generate proton motive force for electron transport phosphorylation. Campylobacter jejuni relies on the use of the amino acids and the citric acid cycle intermediates as carbon sources. Campylobacter jejuni utilize available amino acids in a sequential order, with preference to, serine, aspartate, asparagine, and glutamate and other C. jejuni strains have been found to metabolize proline but only after all other amino acid sources have been completely utilized. Campylobacter can conserve energy via respiration, oxidizing hydrogen and formate for the reduction of electron acceptors such as fumarate, nitrate, sulfites and if at low concentrations, oxygen, to generate proton motive force transport.

There are some C. jejuni subspecies doylei which contain some enzymes from the Entner-Doudoroff pathway which convert glucose-6-phosphate to glyceraldehyde-3-phosphate utilizing gluconate-6-phosphate instead of a fructose-6-phosphate intermediate. This pathway eliminates the need for phosphofructokinase. It has generally been assumed that C. jejuni were found to contain genomic islands (cj0480c-cj0490) that are up regulated in the presence of L-fucose and mucin obtained from the host during colonization in the intestine. The current knowledge of C. jejuni in vivo metabolism is based on amino acid utilization to support the growth and establishment of colonization in the host intestines. Campylobacter jejuni's metabolic diversity is evidenced by its differential carbon utilization.

Campylobacter Pathogenesis and Disease

The exact sequential steps of C. jejuni infection to C. jejuni mediated enteritis are unknown. What is known are the requirements for C. jejuni virulence: motility, drug resistance, host cell adherence, host cell invasion, alteration of the host cell signaling pathways, induction of host cell death, evasion of the host immune system defenses, and acquisition of iron which serves as a micronutrient for growth and works as a catalyst for hydroxyl radical formation. It is also known that C. jejuni secretes proteins that contribute to the ability of the bacterium to invade the host epithelial cells. Survival in the host environment depends on several adaptive responses including adherence, protein secretion, invasion and replication.

The biochemical effects on cellular events are as follows: cytoskeletal rearrangement, host cell death, tight junction disruption and cytokine induction that leads to loss of epithelial cell function, a compromised barrier and absorptive functions, and tissue destruction and disease manifestation. Adherence can also lead to an early inflammatory response that causes stimulation of innate immune functions in the following sequence: an influx of fluid, complement cell activation, recruitment of phagocytes that can lead to C. jejuni lysis or death, clearing of the infection or the presentation of antigens, a humoral response and, a clearing of the infection. In the other direction, instead of C. jejuni lysis or death, the inflammatory response could lead to tissue destruction and disease manifestation.

There is evidence that C. jejuni infections commonly precede Guillain Barré Syndrome which results from a case of molecular mimicry whereby host produced anti-Campylobacter antibodies recognize and cross react with self gangliosides and damage peripheral nerve tissue. Guillain Barré Syndrome is a rare autoimmune disease characterized by the demyelination of motor and sensory nerves or deterioration of axonal nerves of the peripheral nervous system. This nerve damage can lead to muscle weakness, paralysis, and death. It is reported that Guillain Barré Syndrome "has become the most frequent cause of acute flaccid paralysis since the near elimination of poliomyelitis in the world".

Studies have established Guillain Barré Syndrome as a mechanism of molecular mimicry based on Koch's and Witebsky's postulates. They are as follows: the establishment of an epidemiological association between the infectious agent and the autoimmune disease, the identification of T cells or antibodies directed against the patient's target antigens, the identification of microbial mimics of the target antigen, and reproduction of the disease in an animal model. Autoantibodies are the pathogenic component that triggers Guillain Barré Syndrome.

There are two major sub-forms of Guillain Barré Syndrome that affect the peripheral nervous system. The major sub-forms of Guillain Barré Syndrome are: acute inflammatory demyelinating polyneuropathy characterized by the demyelination of peripheral nerves, acute motor axonal neuropathy characterized by degeneration of axonal components of peripheral nerves, and Miller Fischer syndrome characterized by areflexia, ataxia, and ophthalmoplegia. Studies have indicated that C. jejuni infection precedes Guillain Barré Syndrome in 20 to 50% of cases in Europe, North and South America, Japan, and Austraila. The percentage is suspected to be higher in developing countries. Since Campylobacter infections occur far more frequently than Guillain Barré Syndrome, neither the characteristics of the host nor the strain of Campylobacter species are known determinants of which persons with Campylobacter infection contract Guillain Barré Syndrome. Guillain Barré Syndrome usually develops 1 to 3 weeks after a C. jejuni infection. Studies in Japan have indicated that the risk of developing Guillain Barré Syndrome may be higher after infection with C. jejuni type O:19.

Reservoirs of Campylobacter

In developing countries Campylobacter infection is said to be hyperendemic and major sources of infection are environmental and food contamination. Conversely, food production animals are considered to be the primary source of Campylobacter infections in humans in developed countries such as the United States. From animal farms to the commercial production of food commodities there are numerous possibilities for transmission of Campylobacter infection through cross-contamination. Campylobacter infection from consumption of poultry, beef, and pork products is the leading cause of human foodborne illness. Poultry is estimated to account for 50%–70% of human Campylobacter infection. Poultry includes broilers, laying hens, turkeys, ducks, and ostriches. Studies have indicated that the prevalence of Campylobacter colonization in cattle is 0%–80% and 20% in sheep. The same study reported that pigs were more contaminated than cattle or sheep. Poultry and poultry products are considered to be the largest contributor of human Campylobacter infection. Campylobacter jejuni mainly colonizes poultry and is found predominantly in the cecum and colon, but it can also be found in the crops. Horrocks et al. speculated that due to the higher metabolic temperatures of the poultry species they are possibly predisposed to become prominent reservoirs for the thermo tolerant C. jejuni.

Campylobacter transmission is prevalent during preharvest conditions. Colonization in broiler chicks has been demonstrated to be a risk factor in horizontal transmission of C. jejuni infection. Broiler chick colonization is estimated to take place no sooner than seven days of age. Other studies have found that broiler chick colonization takes from zero to three weeks. Colonization of free-ranging chickens is estimated to take place from 0–8 days. Although young animals are very susceptible to colonization by C. jejuni, older broiler chicks closer to processing age have a higher percentage of Campylobacter colonization. Despite horizontal transmission of the flock taking place rapidly, colonization of the flock can take up to several weeks. Studies have estimated that up to 98% in the United States and 60% to 80% in Europe of retail chicken meat is contaminated with C. jejuni. It was found that the skin and giblets have particularly high concentrations of Campylobacter contamination.

Despite the consistent implication that Campylobacter colonization of poultry and poultry products is the major contributor to human foodborne illness, cattle and other swine frequently carry C. jejuni and C. coli. It is probable that cattle carcasses are contaminated during processing either directly or indirectly. None the less, the cases of human foodborne illness as a result of Campylobacter contamination in bovine products (unpasteurized milk and meat) is a legitimate concern.

Multiple studies with cattle have shown that Campylobacter preferentially colonize the lower gastrointestinal tract as opposed to the upper gastrointestinal tract (the 1st stomach) where there is a lower pH environment in the rumen. Additionally the stratified epithelium of the rumen probably lacks the necessary receptors to sustain a persistent high level of Campylobacter colonization. The gallbladder, liver and bile have all been shown to harbor moderately high percentages of Campylobacter. In recent studies, 33% and 21.8% respectively of samples tested positive for Campylobacter in the gallbladder, its mucosal tissue and bile. Unpasteurized bovine milk and milk products are common vehicles for Campylobacter foodborne disease transmission. In a study with cattle from a dairy farm 12% of raw milk samples were found to be contaminated with C. jejuni possibly as a result of contact with bovine feces, contaminated water or direct contamination as a result of mastitis.

The incidence of Campylobacter is reported to be lower for forage fed (primarily dairy cattle) than that of feedlot cattle. The possible contributors to the high incidence of Campylobacter in feedlot cattle are: increased stocking densities, constant contact with feces from other animals, and (3) the high frequency of shared access to community feeding and water troughs.These studies found that there was a higher prevalence of Campylobacter in feedlot cattle (68%) compared to that of adult cattle from the pasture (7.3%) and others found that prevalence rates increased in fed cattle from 1.6% near or upon entry to the feedlot to 63% near the end of the finishing period. Prevalence numbers remained relatively the same whether tests were conducted before or after transport to processing. These results suggest that confinement may promote increased carriage and horizontal transmission of Campylobacter. Further swab tests on the hides of these cattle yielded the same results. The feedlot cattle had significantly more contaminated hides.

The numbers of swine that are colonized by Campylobacter are more comparable to cattle than to those of poultry. Campylobacter coli, which normally inhabit pig intestines, are also found on pork products. Swine are predominantly colonized with C. coli albeit less frequently they are colonized with C. jejuni. Young et al. demonstrates that despite the prevalence of C. coli in swine, a high prevalence of C. jejuni enteric colonization (cecal or rectal contents) has been observed in gilts, sows, and weaned piglets. Studies conducted by Jensen et al. investigated C. jejuni versus C. coli colonization of outdoor organically-reared pigs to monitor possible shifts from C. coli to C. jejuni intestinal colonization. Their findings from three research trials indicated that there were excessive fluctuations (0%, 18.8% and 78.6%) of swine colonized by C. jejuni, and in each trial swine were still dominantly colonized by C. coli. There have also been research studies that demonstrate the possible co-existence of C. coli and C. jejuni in pigs with C. jejuni being present in lower numbers. In these studies; however, C. jejuni was never found to be more prevalent than C. coli.

The life of swine raised for food production starts at a farrowing barn where they are born. After three weeks they are moved to a nursery where they will stay until they reach 22.6 kg (6 weeks). Next they are moved to a finishing unit where they will stay until they reach market growth rate (99.8–113.4 kg). Environmental conditions naturally play an important role in Campylobacter transmission in swine the same as in cattle and poultry. There are increased possibilities for Campylobacter infection and disease transmission in finishing units and breeding farms. There are distinct differences in the prevalence and occurrence of Campylobacter colonization and infection of pigs raised in organic outdoor production systems than those raised in conventional farm systems.

FOODBORNE STAPHYLOCOCCUS

Staphylococcus is a group of bacteria that cause a multitude of diseases. Under a microscope, staphylococcus bacteria are round and bunched together. They can cause illness directly by infection or indirectly through products they make, such as the toxins responsible for food poisoning and toxic shock syndrome.

Staphylococcus Aureus

S. aureus is a commensal and opportunistic pathogen that can cause wide spectrum of infections, from superficial skin infections to severe, and potentially fatal, invasive disease. This ubiquitous bacterium is an important pathogen due to combination of "toxin-mediated virulence, invasiveness, and antibiotic resistance." This organism has emerged as a major pathogen for both nosocomial and community-acquired infections. S. aureus does not form spores but can cause contamination of food products during food preparation and processing. S. aureus can grow in a wide range of temperatures

(7° to 48.5° C; optimum 30 to 37°C), pH (4.2 to 9.3; optimum 7 to 7.5), and sodium chloride concentration up to 15% NaCl. S. aureus is a dessication tolerant organism with the ability to survive in potentially dry and stressful environments, such as the human nose and on skin and inanimate surfaces such as clothing and surfaces. These characteristics favor growth of the organism in many food products. S. aureus can remain viable on hands and environmental surfaces for extended durations after initial contact.

Staphylococcal Food-Borne Disease

SFD is one of the most common FBD and is of major concern in public health programs worldwide. It is one of the most common causes of reported FBD in the United States. The first documented event of SFD due to the consumption of contaminated cheese was investigated by Vaughan and Sternberg in Michigan, USA, in 1884. A typical FBD caused by S. aureus has a rapid onset following ingestion of contaminated food (usually 3–5 hours). This is due to the production of one or more toxins by the bacteria during growth at permissive temperatures. However, the incubation period of SFD depends on amount of toxin ingested. Very small dose of SEs can cause SFD. For example, one report indicated that approximately 0.5 mg/mL concentration of SEs contaminated with chocolate milk caused a large outbreak.

The onset of SFD is abrupt. Symptoms include hypersalivation, nausea, vomiting, and abdominal cramping with or without diarrhea. If significant fluid is lost, physical examination may reveal signs of dehydration and hypotension. Abdominal cramps, nausea, and vomiting are the most common. Although SFD is generally self-limiting and resolves within 24–48 hours of onset, it can be severe, especially in infants, elderly, and immune-compromised patients. Antibiotics are not used for therapy. Approximately 10% of individuals inflicted with SFD will present to a hospital. Management of SFD is supportive. The attack rate of SFD can be up to 85%. S. aureus may not be detected by culture in the events when food is contaminated and toxin is formed prior to cooking. A study involving 7126 cases indicated that case fatality rate of SFD is 0.03%.; all deaths were in elderly patients. Recovery is complete in approximately 20 hours.

The conclusive diagnostic criteria of SFD are based upon the detection of staphylococcal enterotoxins in food, or recovery of at least 105 S. aureus from food remnants. S. aureus enterotoxin can be detected on the basis of three types of methods: bioassays, molecular biology, and/or immunological techniques. Polymerase chain reaction (PCR), reverse transcription PCR (RT-PCR), and RT-quantitative PCR can be carried out to evaluate the toxic potential of strain. The enzyme immunoassay and enzyme-linked fluorescent assay are the most commonly used immunological methods based on the use of antienterotoxin polyclonal or monoclonal antibodies. Several molecular typing methods are widely used for the genetic characterization of S. aureus such as multilocus sequence typing, spa typing, SCCmec typing, and Pulse-field gel electrophoresis (PFGE). These techniques provide means to trace epidemiologically related strains leading to the tracking back to the origin of contamination. However, these methods have variation in their discriminating powers and can be increased by combining the methods. Molecular-based methods provide information about the source of contamination (human or animal origin). The PFGE and spa typing can be used alone or in association to gather the information regarding the origin of S. aureus contamination.

Various types of foods serve as an optimum growth medium for S. aureus. Foods that have been frequently implicated in SFD are meat and meat products, poultry and egg products, milk and dairy products, salads, bakery products, especially cream-filled pastries and cakes, and sandwich fillings. Foods implicated with SFD vary from country to country, particularly due to variation in consumption and food habits. If food is prepared in a central location and widely distributed, SFD outbreaks can have grave consequences impacting thousands of people. For example, over 13,000 cases of SFD occurred in Japan in 2000 as a result of contamination of milk at a dairy-food-production plant.

Staphylococcal Aureus Enterotoxins

S. aureus produces wide arrays of toxins. Staphylococcal enterotoxins (SEs) are a family of nine major serological types of heat stable enterotoxins (SEA, SEB, SEC, SED, SEE, SEG, SEH, SEI, and SEJ) that belong to the large family of pyrogenic toxin superantigens. Pyrogenic toxins cause superantigenic activity such as immunosuppression and nonspecific T-cell proliferation. It is hypothesized that superantigenic activity of SEs helps facilitate transcitosis that allows the toxin to enter the bloodstream, thus enabling it to interact with antigen-presenting cells and T cells leading to superantigen activity. The majority of effects of SEs in SFD is believed to be triggered by initiating a focal intestinal inflammatory response due to their superantigenic activity or by affecting intestinal mast cells causing their degranulation.

SEs are highly stable and highly heat-resistant and resistant to environmental conditions such as freezing and drying. They are also resistant to proteolytic enzymes such as pepsin or trypsin and low pH, enabling them to be fully functional in the gastrointestinal tract after ingestion. The heat stability characteristic of S. aureus imposes a significant threat in food industries. The mechanisms of SEs causing food poisoning are not clearly known. However, it is believed that SEs directly affects intestinal epithelium and vagus nerve causing stimulation of the emetic center. All staphylococcal enterotoxins cause emesis. An estimated 0.1 µg of SEs can cause staphylococcal food poisoning in humans.

SEs produced by some strains of S. aureus are the causative agents of SFD, and SEA is the most common toxin implicated in such events. SEA is highly resistant to proteolytic enzymes. SEA was recovered from 77.8% of all SFD outbreaks in the United States followed by SED (37.5%) and SEB (10%). SEA is the most commonly found enterotoxin among SFD outbreaks in Japan, France, and UK. However, SEC and SEE are also implicated with SFD. The outbreak of gastrointestinal illness via contaminated coleslaw in the United States was caused by SEC produced by methicillin-resistant S. aureus (MRSA) from an asymptomatic food handler. SEC was linked to the SFD outbreak in 1980 in Canada. SEC was also involved in the SFD outbreak during 2001–2003 in Taiwanand 2009 outbreak in Japan. S. aureus is often implicated with caprine mastitis. In sheep, goats, and cattle, SEC was the predominant toxin type detected in S. aureus isolated from mastitis milk. Other studies have documented SEC producers as the most prevalent enterotoxin-producing S. aureus isolated from goat's milkand goat's skin of udder, teats, and milk. Six SFD outbreaks in France in 2009 were caused by SEE present in soft cheese made from unpasteurized milk. Although rare, SEE has also been implicated in the SFD outbreaks in USA and UK. Various new SEs (SEG to SElU2) have been identified. However, only SEH-producing strains have been involved in SFD outbreaks.

S. aureus can survive in multiple host species. Molecular typing such as multilocus sequence typing (MLST) has helped to gain insights about population structure of S. aureus. Studies have identified over 2200 sequence types (STs) of S. aureus using the MLST techniques. The STs can be grouped into clonal complexes (CC). Several studies have indicated that majority of the livestock-associated STs belong to a small number of animal-associated clones. For example, CC97, ST151, CC130, and CC126 are commonly found on bovine infections. CC133 are common among small-ruminants such as sheep or goats. ST1, ST8, CC5, ST 121, and ST398 are found in human host species. ST5 is predominant among poultry isolates. CC133 and ST522 are mostly implicated with mastitis in sheep and goats. One Danish study indicated that ST133 was the predominant lineage in sheep and goats.

References

- Food-poisoning-causes, health: healthline.com, Retrieved 28 April, 2020
- Food-poisoning-caused-by-bacteria-food-toxins, poisoning-from-specific-toxic-agents-to-novel-rapid-and-simplified-techniques-for-analysis: intechopen.com, Retrieved 08 July, 2020
- Diarrhea-causes-1324505: verywellhealth.com, Retrieved 16 June, 2020
- Ian Robert Hall (2003). Edible and Poisonous Mushrooms of the World. Timber Press. p. 103. ISBN 0-88192-586-1
- Bacterial-gastroenteritis, conditions: ada.com, Retrieved 05 August, 2020
- Nausea-and-vomiting, allergies-allergy-symptoms: acaai.org, Retrieved 28 April, 2020
- Ridolo, E; Martignago, I; Senna, G; Ricci, G (October 2016). "Scombroid syndrome: it seems to be fish allergy but... it isn't". Current Opinion in Allergy and Clinical Immunology. 16 (5): 516–21. doi:10.1097/ACI.0000000000000297. PMID 27466827
- Sicherer SH, Sampson HA (February 2014). "Food allergy: Epidemiology, pathogenesis, diagnosis, and treatment". J Allergy Clin Immunol. 133 (2): 291–307, quiz 308. doi:10.1016/j.jaci.2013.11.020. PMID 24388012

Permissions

All chapters in this book are published with permission under the Creative Commons Attribution Share Alike License or equivalent. Every chapter published in this book has been scrutinized by our experts. Their significance has been extensively debated. The topics covered herein carry significant information for a comprehensive understanding. They may even be implemented as practical applications or may be referred to as a beginning point for further studies.

We would like to thank the editorial team for lending their expertise to make the book truly unique. They have played a crucial role in the development of this book. Without their invaluable contributions this book wouldn't have been possible. They have made vital efforts to compile up to date information on the varied aspects of this subject to make this book a valuable addition to the collection of many professionals and students.

This book was conceptualized with the vision of imparting up-to-date and integrated information in this field. To ensure the same, a matchless editorial board was set up. Every individual on the board went through rigorous rounds of assessment to prove their worth. After which they invested a large part of their time researching and compiling the most relevant data for our readers.

The editorial board has been involved in producing this book since its inception. They have spent rigorous hours researching and exploring the diverse topics which have resulted in the successful publishing of this book. They have passed on their knowledge of decades through this book. To expedite this challenging task, the publisher supported the team at every step. A small team of assistant editors was also appointed to further simplify the editing procedure and attain best results for the readers.

Apart from the editorial board, the designing team has also invested a significant amount of their time in understanding the subject and creating the most relevant covers. They scrutinized every image to scout for the most suitable representation of the subject and create an appropriate cover for the book.

The publishing team has been an ardent support to the editorial, designing and production team. Their endless efforts to recruit the best for this project, has resulted in the accomplishment of this book. They are a veteran in the field of academics and their pool of knowledge is as vast as their experience in printing. Their expertise and guidance has proved useful at every step. Their uncompromising quality standards have made this book an exceptional effort. Their encouragement from time to time has been an inspiration for everyone.

The publisher and the editorial board hope that this book will prove to be a valuable piece of knowledge for students, practitioners and scholars across the globe.

Index

www.ingramcontent.com/pod-product-compliance
Lightning Source LLC
LaVergne TN
LVHW080805220926
848485LV00011B/189